Power System Management

Power System Management demonstrates the effectiveness of emerging technologies in electrical systems compared to traditional operational systems. It showcases different operations of the electrical systems, presents the output results based on the latest techniques, and compares the results with classical methods.

- Highlights how to implement the modern automation system in the electrical transmission and distribution systems
- Discusses the integration of distributed energy resources at both medium-voltage (MV) and low-voltage (LV) levels with modern automation systems
- Showcases the problems associated with the AC transmission system and the required automation control system
- Covers application of smart technologies including deep learning and artificial intelligence in fault detection and grid optimization
- Presents real-time monitoring and control of power system devices using the Internet of Things systems, and artificial intelligence-operated robotics system used for control of electrical distribution system

The text is primarily written for senior undergraduates, graduate students, and academic researchers in the fields including electrical engineering, electronics and communications engineering, computer science, and engineering.

Power System Management
Advances and Applications

Edited by
Nagendra Singh, Yogendra Kumar,
Yogesh Pahariya, Vadim Bolshev, and
Harsh Pratap Singh

CRC Press
Taylor & Francis Group
Boca Raton London New York

CRC Press is an imprint of the
Taylor & Francis Group, an **informa** business

Designed cover image: Shutterstock

First edition published 2025
by CRC Press
2385 NW Executive Center Drive, Suite 320, Boca Raton FL 33431

and by CRC Press
4 Park Square, Milton Park, Abingdon, Oxon, OX14 4RN

CRC Press is an imprint of Taylor & Francis Group, LLC

ISBN: 978-1-032-85025-2 (hbk)
ISBN: 978-1-032-85026-9 (pbk)
ISBN: 978-1-003-51615-6 (ebk)

DOI: 10.1201/9781003516156

Typeset in Sabon
by Newgen Publishing UK

Contents

About the editors

Nagendra Singh is working as the Principal of Trinity College of Engineering and Technology, Karimnagar, Telangana, India. He has worked in various positions, like Dean of Academics, HOD, and Exam Superintendent. He has published research articles in reputed refereed Science Cited Indexed and Scopus-indexed journals and conferences. He has published 15 book chapters and edited 5 books, published by Indian and international publishers. He has been granted 14 design patents. He has organized many national and international conferences as well as faculty development programs and seminars. Under his guidance one candidate completed PhD and three candidates completed MTech in Electrical Engineering. His areas of interest are optimization techniques, artificial intelligence (AI), power systems, renewable energy systems, power electronics, smart antennas, and block chain technology. He is a life member of various professional bodies. He is a reviewer for many international journals.

Google scholar Id link: https://scholar.google.com/citations?user=xd7O-GIAAAAJ&hl=en

Yogendra Kumar is working as Professor in the Department of Electrical Engineering, at MANIT Bhopal, MP, India. He has a specialization in power systems. He has 63 articles in reputed SCI, Science Cited Indexed Expanded, and Scopus journals. He has written many articles for international and national conferences. He published so many book chapters in Scopus-indexed books. He guided 19 PhD students and 4scholars in progress. He also guided more than 40 MTech scholars. He has completed many projects and consulting assignments sponsored by different government bodies. He has been granted utility and design patents. He is a reviewer for many international SCI and Scopus journals. His areas of interest are power systems, optimization techniques, AI, and renewable energy systems.

Yogesh Pahariya is presently working as Dean, Academic, Bhagwan Mahavir University, Surat, Gujarat (India) since June 2023. He completed his BE in Electrical Engineering from Shri G. S. Institute of Technology & Science, Indore, in 1991, MTech in Energy Management from School of Energy & Environmental Studies, Devi Ahilyabai University, Indore, in 1994, and PhD from Shri G. S. Institute of Technology & Science, Indore, under Rajeev Gandhi Technological University, Bhopal, in 2011.He has published more than 50 research papers in national and international journals and conferences, which include 2 SCI, 20Scopus, and 10UGC publications. He has published two books and four patents also. He is a fellow member of Institution of Engineers (India), Kolkata and Indian Society of Light Engineers, New Delhi (ISLE), a life member of Indian Society for Technical Education (ISTE), New Delhi, a senior member, International Association of Computer Science and Information Technology (IACSIT), Singapore, and a member, International Association of Engineers, Hong Kong. His research areas are electrical drives, energy auditing, energy management, and power electronics, etc. Pahariya is also certified Energy Manager cum Energy Auditor.

Vadim Bolshev received the MS degree in Engineering from Orel State Agrarian University in 2012 and was a candidate of technical science degree from Federal Scientific Agroengineering Center VIM in 2020. Bolshev gained experience in the industry as an electrician, an electrical engineer, a chief engineer in 2010–2018, and since 2018, he has been working as a researcher in Laboratory of Power and Heat Supply with Federal Scientific Agroengineering Center VIM. In 2021, he headed the Laboratory of Intelligent Agricultural Machines and Complexes with Don State Technical University. The field of scientific activity is to develop technical and economic mechanisms to enhance the efficiency of power supply systems, particularly for rural consumers, and address other issues related to the creation of new engineering technologies for the agro-industrial complex.

Harsh Pratap Singh is currently working as Assistant Professor in Computer Science and Engineering Department at the Shri Vashnav University Indore, India. He obtained his PhD in Computer Science and Engineering from Sri Satya Sai University of Technology and Medical Sciences, Sehore (Madhya Pradesh), India, and he has participated in several conferences. In addition to his academic career, he is a board member of various international journals and worked as a session chair in several conferences. He has published approximately 50 research articles in national/international journals and book chapters.

Contributors

Hemant Amhia
Electrical Engineering Department
Jabalpur Engineering College
India

Neeraj Arya
Department of Computer Science and Engineering
Sri Vaishnav Vidhapeeth Vishwavidyalaya
India

Baidyanath Bag
Department of Electrical Engineering
National Institute of Technology
India

Hari Prasad Bhupathi
Application Software Supervisor
FCA Technology Centre
USA

Vadim Bolshev
Federal Scientific Agroengineering Center VIM
Russia

Alexey Bukreev
Laboratory of Electricity Supply, Electrical Equipment and Renewable Energy
Federal Scientific Agroengineering Center VIM
Russia

Amit Kumar Chandanan
Department of Computer Science and Engineering
Guru Ghasidas Vishwavidyalaya (A Central University)
India

Ashish Choubey
Electrical Engineering Department
Jabalpur Engineering College
India

G.S. Gadge
Department of Electrical Engineering
St. Vincent Pallotti College of Engineering and Technology
India

Lata Gidwani
Department of Electrical Engineering
Rajasthan Technical University
India

Amit Gupta
Department of Electrical and Electronics Engineering
Gyan Ganga Institute of Technology and Sciences
India

Sanjeev Kumar Gupta
Department of Electronics and Communication Engineering
RNTU
India

Bhargavi Jha
Department of Electronics and Communication Engineering
Gyan Ganga Institute of Technology and Sciences
India

Shilpa Kalambe(Jain)
Department of Electrical Engineering
Wainganga College of Engineering and Management
India

Alka Karketta
Department of Electrical and Electronics Engineering
Shri Vaishnav Institute of Technology and Science
India

Sateesh Kourav
Department of Electronics and Communication Engineering
Indian Institute of Information Technology (IIITA)
India

Kiran Kumari
School of Law
Mody University of Science and Technology
India

Neha Smitha Lakra
Department of Electrical Engineering
National Institute of Technology
India

Alexander Lansberg
Junior Researcher
Federal Scientific Agro engineering Center VIM
Russian

Arvind Mewada
School of Computer Science Engineering and Technology
Bennett University
India

Anuprita Mishra
Department of Electrical and Electronics Engineering
IES College of Technology
India

Liza Mozgunova
Queen Mary University of London
United Kingdom

Y. Pahariya
Vice Chancellor
Rai Technological University
India

Ruchi Pandey
Department of Electrical and Electronics Engineering
Gyan Ganga Institute of Technology and Sciences
India

Shriram Pathak
Department of Mechanical Engineering
IES Institute of Technology and Management, IES University
India

Dmitry Pisarev
Laboratory of Electricity Supply, Electrical Equipment and Renewable Energy
Federal Scientific Agroengineering Center VIM
Russia

Deepak Rathod
Department of Electrical and Electronics Engineering
Shri Vaishnav Institute of Technology and Science
India

Rishi Dev
Kalinga Institute of Industrial Technology
India

D.A. Shahakar
Department of Electrical Engineering
P.R. Pote Patil College of Engineering and Management
India

Dipti Shukla
Department of Physics
O P Jindal University
India

Geetam Shukla
Centre for Internet of Things
Madhav Institute of Technology and Science, Deemed University
India

Siddharth Shukla
Department of Electrical and Electronics Engineering
Bansal Institute of Science and Technology
India

Harsh Pratap Singh
Department of Computer Science and Engineering
Shri Vaishnav Vidyapeeth Vishwavidyalaya
India

Kedar Nath Singh
Department of Computer Science and Engineering
Lovely Professional University
India

Maninder Singh
Department of Electronic and Communication Engineering
Motilal Nehru National Institute of Technology Allahabad
India

Nagendra Singh
Department of Electrical Engineering
Trinity College of Engineering & Technology
India

Lokendra Singh Songare
Department of Computer Science and Engineering
Medi-Caps University
India

Anton Sporov
Laboratory of Electricity Supply, Electrical Equipment and Renewable Energy
Federal Scientific Agroengineering Center VIM
Russia

Payal Suhane
Department of Electrical &Electronics Engineering, Vidhyapeeth Institute of Science and Technology
India

M. Sundararajan
Department of Mathematics & Computer Science
Mizoram University
India

A.S. Telang
Department of Electrical Engineering
P.R. Pote Patil College of Engineering and Management
India

Ankit Tiwari
Department of Electrical Engineering
Madhav Institute of Technology & Science
India

Shankarshan Prasad Tiwari
Department of Electrical Engineering
Sandip Institute of Engineering and Management
India

Kirti Verma
Department of Engineering Mathematics
Gyan Ganga Institute of Technology and Sciences
India

Alexander Vinogradov
Federal Scientific Agroengineering Center VIM
Russia

Alina Vinogradova
Federal Scientific Agroengineering Center
Russia

Igor Volynets
Software Development Department
Usetech Bel LLC
Belarus

Prachi Wadhawan
Department of Electrical and Electronics Engineering
Sagar Institute of Research & Technology (SIRT)
India

Leonid Yuferev
Laboratory of Electricity Supply, Electrical Equipment and Renewable Energy
Federal Scientific Agroengineering Center VIM
Russia

Chapter 1

Economic load dispatch by the heuristic optimization techniques

Nagendra Singh, Payal Suhane, Shilpa Kalambe(Jain), Geetam Shukla, and Anuprita Mishra

1.1 INTRODUCTION

Economic load dispatch (ELD) is a technique used to find economic power generation. The objective of the economic dispatch is to minimize the total generation cost while satisfying operational constraints at thermal plants [1]. Minimization of the overall cost of generation while meeting thermal plant operating constraints is the main objective of the economic dispatch [2]. The ELD identifies the load demand and the output power of each available generating unit at any particular time. ELD tries to meet the load demand using the available generation units, and no one generating unit was overloaded at that time [3].

For the optimum solution of the ELD, we can use the various optimization techniques, also called heuristic optimization techniques. In the early days, classical optimization techniques like Lambda iteration, Newton–Raphson, linear programming, etc., were used, but nowadays, heuristic optimization techniques are very effective and efficient for solving nonlinear ELD. With the help of case studies, we can determine the minimum generation cost of the thermal power plant [4].

1.2 ECONOMIC LOAD DISPATCH

ELD is an important optimization task in power system operation for allocating generation among the committed units such that the constraints imposed are satisfied per unit and the generation cost is minimized. Per unit generation cost is defined in terms of British thermal units per hour (Btu/h) or dollars per hour ($/h) [5].

In the literature, various types of ELD models were explained, such as linear ELD, nonlinear ELD, dynamic ELD, ELD with environmental emissions, multi-objective economic dispatch, and ELD with nonconventional energy sources with various constraints [6].

DOI: 10.1201/9781003516156-1

1.2.1 Advantages of ELD

The advantages of ELD are as follows:

- According to load demand, it helps to generate electrical power at the lowest cost.
- Economic dispatch uses the more efficient generation units to minimize the total generation cost.
- It is also sure that the units used for generation need lower fuel consumption, better-utilized fuel, and reduced environmental pollution.
- Economic dispatch helps the operators maintain secure grid operation in such a way as to increase operational reliability without increasing cost.
- ELD policy is very flexible; it encourages the incorporation of the new fuel policy goals and green energy systems to fulfill the respectable demand.
- Economic dispatch will encourage new investment in generation as well as transmission enlargement and upgrades that enhance reliability and cost savings.

1.2.2 The power generation operating cost

The output power of the thermal power plants is increased sequentially by opening a set of valves to their steam turbine at the inlet. The throttling losses are large when a valve is just opened, and they are small when the valve is fully opened [6].

The objective of the ELD is to minimize the generation cost of the thermal power plants. Various types of costs associated with thermal power plants are shown in Figure 1.1. As economics plays a significant role in all fields of industry, the generation power plant is designed in such a way that it operates in the most economical conditions. The primary purpose of the design and operation of the plant is to bring the cost of energy produced to a minimum [7].

1.2.3 Formulation of ELD problem

The main concern with the ELD problem is the minimization of its objective function. To obtain optimum results from the ELD problem, it is necessary to formulate a proper mathematical model with its constraints [8]. A mathematical model of an ELD problem can be formulated in many ways, as listed in the literature, such as a single objective, a multi-objective, a linear model of ELD, a nonlinear model, an ELD model with emission effect, an ELD model with nonconventional sources, etc.

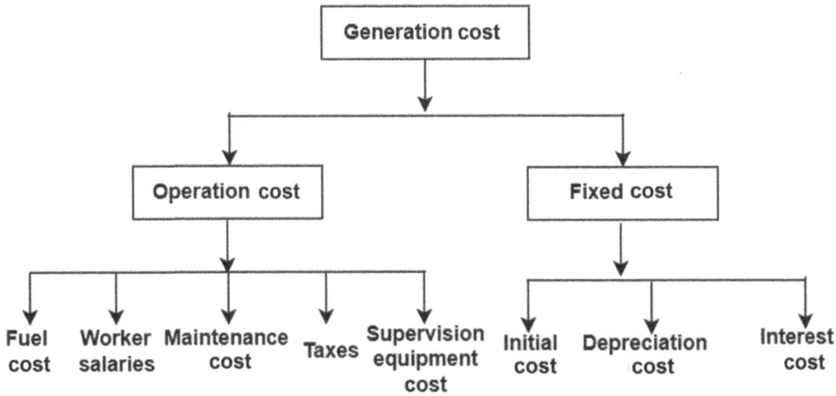

Figure 1.1 Line diagram of the cost associated with power generation plants.

1.2.3.1 Single-objective ELD problems

The main objective of ELD problem is given as

$$\text{Minimize } F_T = \sum_{i=1}^{N} F_i\left(P_i\right), \tag{1.1}$$

where F_T is the total generation cost; F_i is the power generation cost function of the ith unit, and N is the total number of generating units.

1.2.3.1.1 Classical smooth cost function

Generally, the fuel cost of a thermal generating unit is considered as a second-order polynomial function [9]. Figure 1.2 shows the fuel cost curve for the linear ELD problem.

The mathematical model of classical and smooth cost function is given as

$$F_i\left(P_i\right) = a_i P_i^2 + b_i P_i + c_{i,} \tag{1.2}$$

where P_i is the generated power of the ith generating units between the prescribed limit; a_i, b_i, and c_i are the fuel cost coefficients of the ith generating unit.

1.2.3.1.2 ELD problem formulation with valve-point effect

The generating units with multi-valve steam turbines exhibit a greater variation in fuel cost functions [6, 7]. Ripples are induced due to the operation

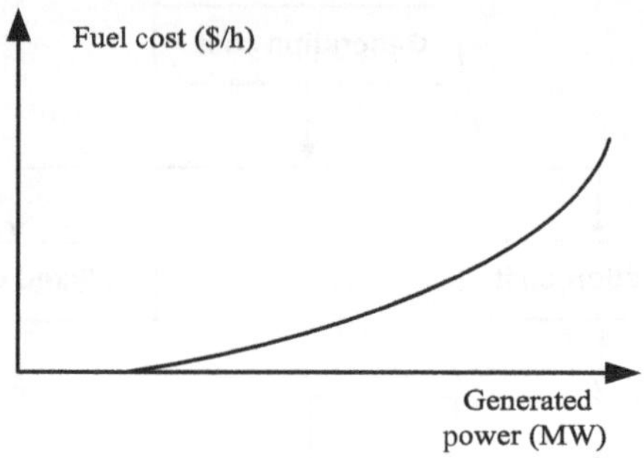

Figure 1.2 Fuel cost curve of thermal generating plants.

of the valve in the plant, and hence the classical fuel cost characteristic is modified considering the valve-point effects [10, 11]. The significance of this effect is that the actual cost curve function of a large steam plant is not continuous. The effects of the operation of the valve are taken into consideration in the ELD problem by superimposing the fundamental quadratic fuel cost characteristics with the rectified sinusoid component [12]. When the valves are operated at thermal plants and ripples are rising, the fuel cost function can be defined with the sinusoidal function of the valve-point effect, given as follows:

$$F_{i,1}(P_i) = F_i(P_i) + \text{abs}(e_i \ \sin(f_i(P_i^{min} - P_i)) \tag{1.3}$$

$$F_{i,1}(P_i) = a_i P_i^2 + b_i P_i + c_i + \text{abs}(e_i \sin(f_i(P_i^{min} - P_i)), \tag{1.4}$$

where $F_{i,1}(P_i)$ is the cost function with valve-point loading effect, and e_i and f_i are the coefficients of ith generator reflecting valve-point operation.

1.2.3.1.3 ELD problem formulation with emissions

Global warming is a concern for the power industry, as it is accountable for the emission of greenhouse gases into the environment. As discussed earlier, the amendment of the Clean Air Act and environment-friendly policies (the Carbon Credit System) develops an interest in the power sector toward reducing environmental emissions (of NO_x, SO_x, and CO_2 gases). Various mathematical model formulations are used to represent the emission of greenhouse gases in the ELD problem [13]. It can be represented in quadratic form by the addition of quadratic polynomials with exponential terms or by the addition of a linear equation with exponential terms of generated power.

The objective function formulated with an emission constraint is given as follows:

$$F_{T1} = F_i(P_i) + k(E_i(P_i)), \tag{1.5}$$

where

$$k = F_{i,Max}/E_{i,Max} \tag{1.6}$$

$$F_{i,Max} = a_i P_{i,Max}^2 + b_i P_{i,Max} + c \tag{1.7}$$

$$E_{i,Max} = d_i P_{i,Max}^2 + e_i P_{i,Max} + f_{i,} \tag{1.8}$$

where F_{T1} is the cost function with emission effect, $E_i(P_i)$ is the emission function, and d_i, e_i, and f_i are the emission coefficients.

1.2.3.2 Multi-objective ELD problem formulation

In recent years, the electricity industry has undergone drastic changes due to a worldwide deregulation or privatization process that has significantly affected the energy market. The bidding mechanism is one of the most challenging issues in the power market. In conventional transaction planning, minimal pool purchase cost or maximal social profit is the market-clearing objective function.

1.2.3.2.1 Objective functions

1.2.3.2.1.1 OBJECTIVE 1

Economic cost function: The main objective of ELD problem, as mentioned in equation (1.9), is taken as the first objective:

$$\text{Minimize } F_T = \sum_{i=1}^{N} F_i(P_i) \tag{1.9}$$

$$F_i(P_i) = a_i P_i^2 + b_i P_i + c_i \tag{1.10}$$

1.2.3.2.1.2 OBJECTIVE II

Emission objective function: The main objective of this study is to minimize total environmental emission cost. A typical NO_x emission [14] can be formulated as

$$\text{Minimize } E_T = \sum_{i=1}^{N} E_i(P_i), \tag{1.11}$$

where

$$E_i(P_i) = (\alpha_i + \beta_i P_i + \gamma_i P_i^2) + \xi_i \sin(\lambda_i P_i). \tag{1.12}$$

Now both objectives may be combined in a single objective as given in equations (1.13), (1.14), and (1.15). Evaluate the cost of each generator at its maximum output:

$$F_i(P_{i.max}) = (a_i P_{i.max}^2 + b_i P_{i.max} + c_i) \tag{1.13}$$

Evaluate the NO_x emission of each generator at its maximum output:

$$E_i(P_{i.max}) = (\alpha_i + \beta_i P_{i.max} + \gamma_i P_{i.max}^2) + \xi_i \sin(\lambda_i P_{i.max}) \tag{1.14}$$

Divide the cost of each generator by its NO_x emission:

$$\frac{F_i(P_{i.max})}{E_i(P_{i.max})} = k_i, \tag{1.15}$$

where E_T is the total emission for i^{th} generating units, $E_i(P_i)$ is the emission function for i^{th} generating units, and α_i, β_i, γ_i, ξ_i, and λ_i are the environmental emission coefficients.

1.2.3.3 Constraints

1.2.3.3.1 Power balance constraint

The classical and non-classical models, either with smooth or non-smoothed fuel cost functions, are subjected to the following equality and inequality constraints [15]:

$$\sum_{i=1}^{n} P_i = P_D + P_L, \tag{1.16}$$

where P_D is the total demand and P_L is the line loss.

1.2.3.3.2 Generator power limits

Each generator is constrained between its minimum and maximum limits, as shown in equation (1.17):

$$P_i^{min} \leq P_i \leq P_i^{max}, \tag{1.17}$$

where P_i is the output power of ith generator; P_i^{min} and P_i^{max} are the min/max power outputs of ith generator.

1.2.3.3.3 Ramp-rate limit constraint

The actual operating ranges of all online units are restricted by their corresponding ramp-rate limits [16]. As the operating points increase and decrease of the generators, ramp-rate limit constraints are described as follows:

1. As generation increases

$$P_i(t) + P_i(t-1) \leq UR_i \tag{1.18}$$

2. As generation decreases

$$P_i(t-1) - P_i(t) \geq DR_i \tag{1.19}$$

When the generator ramp-rate limits are considered, the operating limits for each unit, output is limited by the time-dependent ramp up/down rate at each hour as given below:

$$P_i^{min}(t) = \max\left(P_i^{min}, P_i(t-1) - DR_i\right) \tag{1.20}$$

$$P_i^{max}(t) = \min\left(P_i^{max}, P_i(t-1) - UR_i\right) \tag{1.21}$$

$$P_i^{min}(t) \leq P_i(t) \leq P_i^{max}(t) \tag{1.22}$$

1.2.3.3.4 Prohibited operating zone

The generating units may have certain ranges where operation is restricted on the grounds of the physical limitations of machine components or instability. These restrictions may be due to the operation of steam valves or vibration in shaft bearings [17]. Consequently, if these prohibited operating ranges are incorporated into the cost curve of the thermal generating units, then it leads to discontinuities in the cost curves that need to avoid operations in these zones to obtain a feasible economic solution:

$$P_i \in \begin{cases} P_i^{min} \le P_i \le P_{i1}^{L} \\ P_i^{Uk-1} \le P_i \le P_i^{Lk}, \quad k = 2,3,\ldots Zi, \\ P_{iZi}^{Uk} \le P_i \le P_i^{max}, \\ i = 1,2,\ldots PZ, \end{cases} \tag{1.23}$$

where Z_i are is the number of prohibited zones in the ith generator curve, k is the index of prohibited zones of ith generator, P_i^{Lk} is the lower limit of kth prohibited zone, and P_i^{Uk-1} is the upper limit of the kth prohibited zone of the ith generator. PZ is the number of generators with prohibited operating zones.

1.2.3.3.5 Spinning reserve constraints

Spinning reserve in a thermal power system is a relevant economic issue in a deregulated environment. For maintaining the reliable system, sufficient spinning reserves are required [18], given as equation (1.24):

$$\sum_{i=1}^{N} P_i^{max} \ge P_D + P_L + R), \tag{1.24}$$

where R is the value of spinning reserve.

A fixed amount, a predefined percentage of maximum load demand, or the size of the single largest generator is used as a spinning reserve. A higher spinning reserve makes the system more reliable; however, it will increase running costs. Mainly, spinning reserve is fulfilled in unit commitment scheduling.

1.2.3.3.6 Security constraints

A power system is normally designed to be secure for any simple contingency involving the outage of a transmission line, a generator, and a load [19]. If any one such contingencies occurs, then there will normally be no violations of its constraints. When more than one contingency is involved, violations of equality and inequality constraints need to be examined.

1.2.3.3.7 Network losses

In the ELD, the network losses should be taken into account as functions of the respective generating unit output using B coefficients [20], which are given as

$$P_L = \sum_{i=1}^{n}\sum_{j=1}^{n} P_i B_{ij} P_j + \sum_{i=1}^{n} B_{i0} P_i, \tag{1.25}$$

Where B_{ij} is the ij[th] element of loss coefficient symmetric matrix B, B_{i0} is the element of the loss coefficient vector, and $B_{00}$is the loss coefficient constant.

1.3 HEURISTIC OPTIMIZATION TECHNIQUES

Recent techniques that are efficient in computation and have the ability to search for a better solution to complex optimization problems have fostered the development of a group of techniques known as evolutionary programming techniques [21, 22], such as simulated annealing, genetic algorithms, artificial neural network (ANN), tabu search, fuzzy logic, ant colony, differential algorithms, evolutionary programming, and particle swarm optimization. Heuristic programming techniques are based on stochastic optimization, which is able to search for solutions to problems using a simplified model of the evolutionary process found in nature. These algorithms provide an alternative for obtaining global optimal solutions to the problems, which are discontinuous, non-convex, and highly nonlinear in nature.

Heuristic programming seeks the optimal solution to an optimization problem by evolving a population of candidate solutions over a number of generations or iterations. A new population is formed from an existing population using a mutation operator. This operator perturbs each component of every solution in the population by a random amount to produce innovative solutions.

Heuristic techniques are robust optimization techniques that are suitable for finding the optimum solution to discontinuous, nonlinear, and multi-objective problems. Optimization is a task in a problem to find the most suitable solution. The optimization process starts randomly in the search space, finds the solution to the problem by using the objective function and constraints of such problems to drive its search toward the feasible region, and finally finds the solution to the problem that is near the optimum solution by exploring as small a set of solutions as possible.

The heuristic technique has emerged as a useful optimization tool for handling nonlinear programming problems [22].In real-world problems, it can implement algorithms without complex procedures and does not require extensive modifications to the software to handle additional constraints. In evolutionary techniques, the search process is parallel; for this reason, it has a high probability of finding the optimal solution to the problems [15].

Here we are discussing one of the most effective and popular heuristic optimization techniques, called particle swarm optimization (PSO).

1.3.1 Particle swarm optimization

In the optimization of a design, the objective could be simply to minimize the cost or to maximize the efficiency of the production. An optimization technique is a procedure that is executed iteratively by comparing various solutions until an optimum or satisfactory solution is found.

The PSO method was introduced in 1995 [23]. This method is motivated by the social behavior of organisms. It is a population-based search technique. In the PSO technique, particles are initialized randomly in the search space and change their position with time. These particles fly around in a multidimensional problem-search area. When these particles move in the search area, they adjust their position according to their own experience as well as the experience of neighboring particles. Thus, each particle makes use of the best position encountered by itself and its neighbors. The direction of the particle is given by the set of particles neighboring the particle and its past experience.

PSO is a computational intelligence-based technique that is not largely affected by the size and nonlinear nature of the problems. It is able to provide an optimal solution to the problem where analytical methods fail to converge [19]. It can, therefore, be effectively applied to different optimization problems in power systems.

1.3.1.1 Basic concepts of PSO algorithm

PSO is based on two fundamental disciplines: social science and computer science. In addition, PSO uses the swarm intelligence concept, which is the property of a structure. The collective behaviors of agents that are interacting locally with their environment produce logical globally efficient patterns.

1.3.1.2 Social concept

It is known that human intelligence results from social relationships. Humans learn many things through the valuation, assessment, and reproduction of others, as well as from experience and adapting to the environment. They can determine optimal patterns of behavior, attitudes, and whatever. Another basic social concept indicates that civilization and cognition are inseparable consequences of human sociality.

1.3.1.3 Swarm intelligence principles

Fundamental principles of swarm intelligence can be described by considering five terms:

1. *Proximity principle:* The population should be able to carry out simple space and time computations.

2. *Quality principle:* The population should be able to respond to quality factors in the environment.
3. *Diverse response principle:* The population should not commit its activity along excessively narrow channels.
4. *Stability principle:* The population should not change its mode of behavior every time.
5. *Adaptability principle:* The population should be able to change its mode of behavior. When it is worth the computational value.

In PSO, the term "particles" refers to population members that are mass-less and volume-less (or with an arbitrarily small mass or volume) and are subject to velocities and accelerations toward a better mode of behavior.

1.3.1.4 Computational characteristics

Swarm intelligence provides a useful paradigm for implementing adaptive systems [24]. It is an extension of evolutionary computation and includes the softening parameterization of logical operators such as AND, NOT, and OR.

1.3.1.5 Representation of the swarm

As the decision variables for the ELD problems are real power generations, they are used to form the swarm. The real power output of all generators is represented by the positions of the particles in the swarm. For a d-dimensional problem, particles are represented as follows:

$$P_S = \left(P_{i1}, P_{i2}, \ldots\ldots\ldots P_{id}\right), \tag{1.26}$$

where P_S is the set of position vector of the ith particle, it represents one of the possible solutions for the optimization problem.

Similarly, select the velocity of the particles according to the limits of power generation, which is represented as:

$$V_i^K = \left(V_{i1}, V_{i2}, \ldots\ldots\ldots V_{id}\right), \tag{1.27}$$

where V_i^K is the initial velocity of the particles in the d-dimension search space.

1.3.1.6 Initialization of the best positions

In the strategy of PSO, the particle's best position and the global best position are the key factors. The best position of a particle is the position that gives the minimum value of the generated power of the generating units, and the best position out of all the pbest is taken as the gbest.

1.3.1.7 Parameter selection for particle swarm

When implementing the particle swarm algorithm, several considerations must be taken into account to facilitate convergence and prevent an "explosion" of the swarm. These considerations include limiting the maximum velocity, selecting acceleration constants, and determining the inertia constant.

1.3.1.8 Selection of maximum velocity

At each iteration step, the algorithm proceeds by adjusting the distance (velocity) that each particle moves in every dimension of the problem hyperspace. The velocity of the particle is a stochastic variable and is, therefore, subject to creating an uncontrolled trajectory, making the particle follow wider cycles in the problem space [15]. To dampen these oscillations, upper and lower limits can be defined for the velocity:

$$\text{If } V_i > V_{Max} \text{ then } V_i = V_{Max} \text{ else if } V_i < -V_{Max} \text{ then } V_i = -V_{Max}$$

Most of the time, the value is selected according to the characteristics of the problem. It is important to note that if the value of this parameter is too high, then the particles may move erratically, going beyond a good solution; on the other hand, if the value is too small, then the particle's movement is limited and the optimal solution may not be reached.

1.3.1.9 Selection of acceleration constant

Acceleration constants c1 and c2 control the movement of each particle toward its individual and global best positions, respectively. Small values limit the movement of the particles, while large numbers may cause the particles to diverge. The effect of considering a random value for the acceleration constant helps to create an uneven cycle for the trajectory of the particle when it is searching around the optimal value. In the literature, most of the authors selected values of c1 and c2 = 2 and obtained the solution to the problem.

1.3.1.10 Selection of inertia weight

The inertia constant can be either implemented as a fixed value or can be dynamically changed [12]. Essentially, this parameter controls the exploration of the search space; therefore, an initially higher value (typically 0.9) allows the particles to move freely to find the global optimum neighborhood quickly. Once the optimal region is found, the value of the inertia weight can be decreased (usually to 0.4) to narrow the search, shifting from an exploratory mode to an exploitative mode. It is represented as follows:

$$W = W_{max} - \frac{W_{max} - W_{min}}{iter_{max}} \times iter, \tag{1.28}$$

where W is the inertia weight constant.

1.3.1.11 Update the velocity and position of the particles

The particles in the swarm are moved to new positions with the help of new velocities. The velocity and the position of the ith particle are updated as follows:

$$V_i^{(K+1)} = WV_i^K + c_1 Rand_1 \times \left(Pbest_i - S_i^K\right) + c_2 Rand_2 \times \left(gbest - S_i^K\right) \tag{1.29}$$

$$S_i^{(K+1)} = S_i^K + V_i^{K+1}, \tag{1.30}$$

where pbest = (pbest, pbest,..., pbest) is the best previous position yielding the best fitness value for the ith particle, and gbest = (gbest, gbest,..., gbest) is the best position discovered by the whole population. c_1 and c_2 are the acceleration constants reflecting the weighting of stochastic acceleration terms that pull each particle toward the pbest and gbest positions, respectively. $rand_1$ and $rand_2$ are two random numbers in the range [0, 1].

1.3.1.12 Updating the best positions

The best positions of the particles are evaluated with the new positions of the particles. The best position of the particles out of all the new particles is taken as the global best position of the particles (gbest).

1.3.1.13 Stopping criterion

The maximum number of iterations is taken as the stopping criterion.

1.4 CONCLUSIONS

ELD helps find the minimum operating cost of the thermal power plants. As the operating cost depends on the connected load, the load varies throughout the day. Therefore, it is very difficult to find the optimum value of the generation cost. Heuristic optimization techniques like genetic algorithms, PSO, etc. help optimize ELD and evaluate the minimum generation cost. Such optimization techniques are very effective and can be used to solve single- or multi-objective linear and nonlinear ELD problems. The computation time of such techniques is also very short, so they can solve large amounts of data within seconds. Many researchers used such techniques to solve the various case studies of ELD.

REFERENCES

1. Singh N, Chakrabarti T, Chakrabarti P, Panchenko V, Budnikov D, Yudaev I, Bolshev V. Analysis of Heuristic Optimization Technique Solutions for Combined Heat-Power Economic Load Dispatch. *Applied Sciences*. 2023; 13(18):10380. https://doi.org/10.3390/app131810380
2. Singh N, Chakrabarti T, Chakrabarti P, Margala M, Gupta A, Krishnan SB, Unhelkar B. A New PSO Technique Used for the Optimization of Multiobjective Economic Emission Dispatch. *Electronics*.2023; 12(13):2960. https://doi.org/10.3390/electronics12132960
3. Singh N, Chakrabarti T, Chakrabarti P, Margala M, Gupta A, Praveen SP, Krishnan SB, Unhelkar B. Novel Heuristic Optimization Technique to Solve Economic Load Dispatch and Economic Emission Load Dispatch Problems. *Electronics*. 2023; 12(13):2921. https://doi.org/10.3390/electronics12132921
4. Singh N, Sharma AK, Tiwari M, Jasiński M, Leonowicz Z, Rusek S, Gono R. Robust Control of SEDCM by Fuzzy-PSO. *Electronics*. 2023; 12(2):335. https://doi.org/10.3390/electronics12020335
5. Singh N, Nighoskar A, Ram Krishna S. Economic Load Dispatch with Valve Loading Effects Optimize by BBMO. *International Journal of Innovative Technology and Exploring Engineering (IJITEE)*. 2019; 8(7):104–107.
6. Singh N, Praveen Kumar M, Santosh Kumar B. Effect of Valve Loading on the Thermal Power Economic Load Dispatch Using New Elephant Herding Optimization. *International Journal of Recent Technology and Engineering (IJRTE)*. 2019; 7(6S4):345–349.
7. Singh N. A New PSO Techniques for the Optimization of Economic Load Dispatch of Thermal Generated Power. *International Journal of Engineering & Technology*. 2018; 7(16):73–76.
8. Singh N. Renewable Energy for Sustainable Rural Development and Challenges. *IJRAR*. 2019; 6(1):185–187.
9. Singh N, Nagar H, Kumar Y. New Particle Swarm Optimization to Evaluate Solution of ELD Problem. *International Journal of Advanced Technology & Engineering Research (IJATER)*. 2015; 5(6):40–51.
10. Almas K, Singh N. Review on Power Quality Improvement for Grid Connected Wind Energy Generation System. *International Journal of Grid and Distributed Computing*. 2020; 13(1s):106–112. http://sersc.org/journals/index.php/IJGDC/issue/view/294
11. Singh N, Kumar Y. Multi Objective Economic Load Dispatch Problem Solved by New PSO. *Advances in Electrical Engineering*. 2015; 2015:2356–6655.

12. Singh N, Kumar Y. Economic Load Dispatch with Valve Point Loading Effect and Generator Ramp Rate Limits Constraint Using MRPSO. *International Journal of Advanced Research in Computer Engineering & Technology (IJARCET)*. 2013; 2(4):1472–1473.
13. Singh N, Kumar Y. Moderate Random PSO Using for Economic Load Dispatch. *International Journal of Engineering Research & Technology (IJERT)*. 2012; 1(7): 1–7.
14. Singh N, Tamrakar S, Mewada A, Gupta SK. *Artificial Intelligence Techniques in Power Systems Operations and Analysis* (1st ed., pp. 1–72), 2023, Auerbach Publications. https://doi.org/10.1201/9781003301820
15. Kumawat M, Singh N. *Emerging Trends in Science & Technology*, 2021. Bhumi Publisher, ISBN: 978-81-951982-2-1.
16. Singh N, Mishra A, Gupta M, Dwivedi K, Swarnkar M. Implementation of Microgrid for Rural Electrification to Cater on Side Load Demand. In: *3rd (IEEE) International Conference on Energy, Power and Electrical Engineering (EPEE 2023), Huazhong University of Science and Technology, Wuhan, China*, 2023, pp. 1058–1061. doi: 10.1109/EPEE59859.2023.10351810. The conference has organized a special session for Indian participants on 16 September2023, at venue IES University Bhopal, India. https://ieeexplore.ieee.org/document/10351810
17. Singh N, Kumar Y. Analysis of Optimum Cost and Size of the Hybrid Power Generation System Using Optimization Technique. In: *12th IEEE International Conference CSNT-2023 held at Technocrats Institute of Technology (Excellence) Bhopal, Madhya Pradesh (India)*, 8 and 9 April 2023, pp. 284–291, doi: 10.1109/CSNT57126.2023.10134720.. https://ieeexplore.ieee.org/document/10134720
18. Singh N, Kumar Y. Economic Load Dispatch Using MRPSO with Generator Ramp Rate Limits Constraint. In: *Fourth IEEE International Conference on Computational Intelligence and Communication Networks, Mathura, Uttar Pradesh*, pp. 631–637, 3–5 November 2012. doi: 10.1109/CICN.2012.101; ISBN: 978-0-7695-4850-0/12.
19. Singh N, Kumar Y. Economic Load Dispatch with Environmental Emission Used MRPSO. In: *3rd IEEE International Advance Computing Conference (IACC), Ghaziabad, Uttar Pradesh*, pp. 986–990, 22 and 23 February 2013, pp. 995–999, doi: 10.1109/IAdCC.2013.6514362. ISBN: 978-1-4673-4529-3/12.
20. Singh N, Tirole R. Bumble Bees Mating Optimization Algorithm for Economic Load Dispatch with Pollution. In: *IEEE International Conference on Advanced Computation and Telecommunication, ICACAT-2018, Lakshmi Narayan College of Engineering and Technology Excellence, Bhopal, Madhya Pradesh*, 28–29 December 2018, pp. 1–6, doi: 10.1109/ICACAT.2018.8933681. https://ieeexplore.ieee.org/document/8933681
21. Singh N, Kumar Y, Nigoskar A. Optimization of Economic Dispatch Using PSO with Time Varying Acceleration Coefficient. In: *Second IEEE International Conference on Data, Engineering and Application 2020, Organized by School of Information Technology RGPV, Bhopal, Madhya Pradesh*, 28–29 February 2020, pp. 1–4, doi: 10.1109/IDEA49133.2020.9170662. ISBN:978-1-7281-5719-1. https://ieeexplore.ieee.org/document/9170662

22. Singh N, Priyadarshi N, Kumar N. Minimization of Environmental Emission and Cost of Generation Using Economic Load Dispatch with Optimization Techniques. *AIP Conference Proceedings*, (vol. 2771, no. 1), 2023. AIP Publishing.
23. Sondhiya V, Dwivedi K, Kulsum Almas S, Singh N. Energy Management of Nanogrid through Flair of Deep Learning from IoT Environments. In: *Artificial Intelligence Techniques in Power Systems Operations and Analysis*, vol.1, pp.20–35, 2023, Taylor& Francis. ISBN: 9781003301820.
24. Singh N. Optimization Techniques Used for Design of Economic Electrical Power Distribution. In: *Meta Heuristic Optimization Techniques*, pp.79–92, 2022, Walter De Gruyter GmbH, Berlin, Germany. https://doi.org/10.1515/9783110716214-005

Chapter 2

Design of electrically safe physical model of 0.4 kV electric network

Alexander Lansberg, Alexander Vinogradov, and Vadim Bolshev

2.1 INTRODUCTION

A large number of works are aimed at modeling electrical networks, which is one of the most reliable ways to study the processes occurring in them. At the same time, computer modeling is used in most works. So, in work [1,2], fault modeling was carried out in the MATLAB/Simulink software package using the example of a distribution network operating in Nigeria, powered by a two-transformer 132/33 kV substation with a unit transformer power of 30 MVA to assess its reliability in different operating modes. It has been determined that with a two-phase short circuit, the reliability indices of the distribution network SAIDI and SAIFI are, respectively, 1.125 and 8 p.u., with a three-phase short circuit—2.7 and 4 p.u., and with a single-phase ground fault—3 and 5 p.u.

A scientific team from of South American scientists [3] carried out a study of the influence of higher harmonics and asymmetry on electric motor operation. Using the example of an electric motor with a power of 37.3 kW, it was revealed that due to the presence of higher harmonics, the efficiency decreases by 2%–4% due to additional arisen losses.

The research [4] presents the analysis of power quality performed in in the MATLAB/Simulink program using the example of a distribution network with a frequency of 50 Hz, powered by a power system with a voltage of 33 kV and a power of 50 MVA. The work examines the influence of the switching processes for various types of loads of 11 kV side on the sinusoidal shape of the voltage curve and the value of phase voltages and provides oscillograms of the modes.

The interesting work of electrical network modeling is the assessment of the single-phase short-circuit current at the most remote point of the 0.4 kV electrical network, carried out in article [5] at the test site of PJSC "Rosseti Ural"-"Chelyabenergo", Russia. The experimental network selected for testing has a short length (481 m) and includes two types of wire: aerial bundled cable 3×30+1×50 mm^2 and reinforced aluminum-cable steel 35 mm^2. The network under consideration was modeled in MATLAB

DOI: 10.1201/9781003516156-2

Simulink [6], but it has a disadvantage. The model contains only two measuring units at both the short-circuit point and the terminals of the 0.4 kV step-down transformer. It does not make it possible to assess the voltage distribution along the line under established emergency asymmetric modes.

At the same time, a similar physical model developed in research [7] is characterized by the same disadvantages as the mathematical one, and is also not visual and does not meet the electrical safety requirements when simulating emergency modes due to its power supply from power transformer380/177 V with power of 0.5 kVA. The specified physical model is also not universal. It does not allow simulating different types of supply transformers, power line conductors, does not provide the ability to connect different types of loads, and is also not suitable for simulating the parallel operation of renewable energy sources with centralized power supplies, modeling the operation of a second power source, including a single-phase one as a diesel generator. Also relevant is the study of electrical safety issues and the possibility of using means to increase power supply reliability and their impact on power quality, which can also be studied using a physical model.

The second method also allows determining the value of the single-phase short-circuit current by algebraically adding the impedance of the elements up to the short-circuit point and is called the simplified "phase-zero" loop method [8], while its value is no more than 5% less than that obtained results by the method of symmetrical components. Other methods for determining short-circuit current are computer modeling, for example, using the MATLAB Simulink program [9] or the mathematical model developed by the commercial company Project RZA [10]. The models allow not only to determine the value of the single-phase short-circuit current, but also to study other electrical parameters: voltage distribution along the line, voltages and currents of direct, negative and zero sequences at different points of the network. At the same time, the results obtained from them are 25% lower than the calculations using the first two methods.

It should be noted that there are also practical methods for determining the single-phase short-circuit current used in power grid companies: using a device designed to measure the resistance of the phase-zero loop and calculate the expected short-circuit current at the current phase voltage, as well as direct measurement of the emergency current mode. At the same time, these measures are not currently being implemented in power grid companies due to the lack of financial resources for their implementation. In this regard, there is a need for physical modeling and determination of the most reliable method for estimating the single-phase short-circuit current. In this regard, this paper presents the developed physical model of a 0.4 kV electrical network, justifies its parameters, and evaluates its reliability.

The purpose of the study is to develop, justify the parameters, and evaluate the reliability of a physical model of a 0.4 kV electrical network containing a power line sectionalizer and a renewable energy source.

2.2 MATERIALS AND METHODS

The design of the physical model was carried out based on the configuration of the existing 0.4 kV electrical network of the Orel region, Russia, with the dispatch name of 0.4 kV overhead power transmission line No. 2, 10/0.4 kV transformer substation Pishch-922-1, 10 kV overhead line No. 922, power substation "Pishchevaya" 110/10 kV, presented in [11]. For the physical model, a configuration similar to the specified line was provided with a trunk power transmission line (PTL) and a branch line from the trunk one. The length of the trunk PTL and branch lines to the most remote ones is more than 500 m from the transformer substation (TS) 6-10/0.4 kV. It is more than the recommendations according to rules [12].These values are due to the practical features of 0.4 kV electrical networks. For example, in the branches of power grid companies PJSC "Rosseti Center" and PJSC "Rosseti Center and Volga Region" the largest and smallest average lengths in the Kirov and Orel regions, rangingfrom737 to 985 m [13].

The parameters were justified in accordance with the provisions presented in [14, 15], which allows studying the physical processes of the existing electrical network by reducing power, voltages, and currents, and subsequently, using similarity coefficients, to obtain the characteristics of the original by recalculation.

A step-down power transformer with an oil natural air natural cooling system with a capacity of 250 kVA of voltage class 10/0.4 kV and a circuit and a winding connection group Y/Yn-0 is used as a power source (ONAN-250-10/0.4 kV Y/Yn-0). Such transformer is 13% out of the total number used in the power grid company branch of PJSC "Rosseti Center"–"Orelenergo". Calculations were also carried out for transformers of types ONAN-63, ONAN-100, andONAN-160, constituting 14%, 23%, and 20%, respectively, of the electric grid organization fleet [16].

The characteristics of the specified types of transformers that were used in the calculation of similarity coefficients are presented in Table 2.1 [17].

Table 2.1 Technical characteristics of 6–10 kV power transformers

Transformers of ONAN design							
Power, S, kVA	*Combination voltage, U_I/U_2, κβ*		*Short-circuit loss, P_k, W*	*Short-circuit voltage, U_k, %*	*Active resistance, R_t, ohm*	*Reactance, X_t, ohm*	*Impedance, Z_t, ohm*
	HV	*LV*					
63	6–10	0.4	1280	4.5	0.05160	0.114	0.125
100	6–10	0.4	2000	4.5	0.03200	0.072	0.079
160	6–10	0.4	2600	4.5	0.01625	0.045	0.048
250	6–10	0.4	3700	4.5	0.00947	0.029	0.030

Table 2.2 Technical characteristics of transformer TT-0.16-380/38 V

Specifications	*Transformer TT-0.16-380/38 V*
Short-circuit loss, P_{SC}, W	10
Short-circuit voltage, U_{SC}, %	3.7
Rated phase current on the high-voltage side, $I_{rated.HV}$, A	0.243
Rated phase current on the low-voltage side, $I_{rated.LV}$, A	2.43
Active resistance, R_t, ohm	0.56
Reactance, X_t, ohm	0.33
Impedance, Z_t, ohm	0.65

The developed physical model of a 0.4 kV electrical network has a voltage class of 380/38 V, which corresponds to a voltage transformation ratio of 10 p.u. To power the physical model, a three-phase low-power transformer of type TT-0.16-380/38 V with natural air cooling (dry) was used with a Y/Yn-0 winding connection circuit. To determine the technical characteristics of the transformer TT-0.16-380/38 V, a short-circuit experiment was implemented and its technical characteristics were determined, which will be used in calculating similarity coefficients. The obtained technical characteristics of the transformer TT-0.16-380/38 V are presented in Table 2.2.

The methodology for determining the similarity coefficients of the physical model necessary for calculating other parameters of the 38 V physical model was carried out using the basic principles of similarity theory [14]. Initially, similarity coefficients were determined that relate to both voltages of the existing network and the model as well as the impedances of the transformers. The voltage similarity coefficient is determined by equation (2.1) [14]:

The voltage similarity coefficient is determined by equation (2.1) [14]:

$$k_U = \frac{U_m}{U_o}(\mathrm{pu}), \tag{2.1}$$

where U_m is the rated voltage of the physical model of electrical network, V; U_o is voltage of the current electrical network (original, prototype model), V.

The similarity coefficient for the transformer resistance is determined by equation (2.2) [14]:

$$k_Z = \frac{Z_m}{Z_o}(\mathrm{pu}), \tag{2.2}$$

where Z_m is the total resistance of the low-power transformer TT-0.16-380/38 V, supplying power to the physical model, ohm; Z_o is the total resistance of the prototype industrial transformer (in the case under consideration, ONAN-250-10/0.4 kV Y/Y_n), ohm.

The current similarity coefficient is determined by equation (2.3) [14]:

$$k_I = \frac{k_u}{k_z}(pu), \tag{2.3}$$

where m_u is the voltage similarity coefficient, p.u.; m_z is the similarity coefficient in terms of transformer resistance, p.u.

The power similarity coefficient is determined by equation (2.4) [14]:

$$k_S = \frac{k_u^{\ 2}}{k_z}(pu), \tag{2.4}$$

Similarity coefficients allow simulating any type of wire with a transformer of a given power. Moreover, for transformers of different rated powers the same length of a wire of a given PTL crosssection (e.g., 1 km) on the physical model will differ significantly from each other due to the fact that the similarity coefficients are determined by the similarity of the transformers.

To simulate sections of power transmission lines on a physical model, nichrome wire made of X20N80 alloy with a cross section of 0.5 $mm^2$is used, the resistance of which is 5.6 ohm/m [18].

Let's consider the determination of similarity coefficients for the aluminum wire of 35 mm^2 (A-35), which in the Orel power system carries up to 27% out of the total length of 0.4 kV overhead PTLs, while in individual lines its length can reach 100% [16]. Resistance of 1 km line when using A-35 wire is determined by equations (5–7) [19]:

$$x_L = x_{spec} \cdot L \tag{2.5}$$

$$r_L = r_{spec} \cdot L \tag{2.6}$$

$$z_L = \sqrt{r_L^{\ 2} + x_L^{\ 2}} \tag{2.7}$$

where L is PTL length, km; x_L is specific inductive resistance of PTL, 83 ohm/km for A-35 wire [19]; r_L is specific active resistance of PTL, A-35 0.352 ohm/km for A-35 wire [19]; z_L is specific PTL impedance, ohm/km.

Next, using the obtained resistance similarity coefficients, the length of this wire is determined on the physical model. The load factor of a power transformer when modeling emergency modes and using any type of load is determined by equation (2.8) [14]:

$$\beta_7 = \frac{I_{\text{phase(m)}}}{I_{\text{rated.LV}}} \cdot 100, \%, \tag{2.8}$$

where $I_{\text{phase(m)}}$ is the phase current of the model in the operating mode under study, A; $I_{\text{rated.LV}}$ is rated current of the low-voltage winding of the transformer TT-0.16-380/38V, A.

An electrical circuit was developed to implement the physical model of a 0.4 kV electrical network, shown in Figure 2.1. The circuit contains the following: QF1 is input four-pole circuit breaker; QF2, QF3 are three-pole circuit breakers; QF4, QF5 are two-pole circuit breakers, QF6 is single-pole

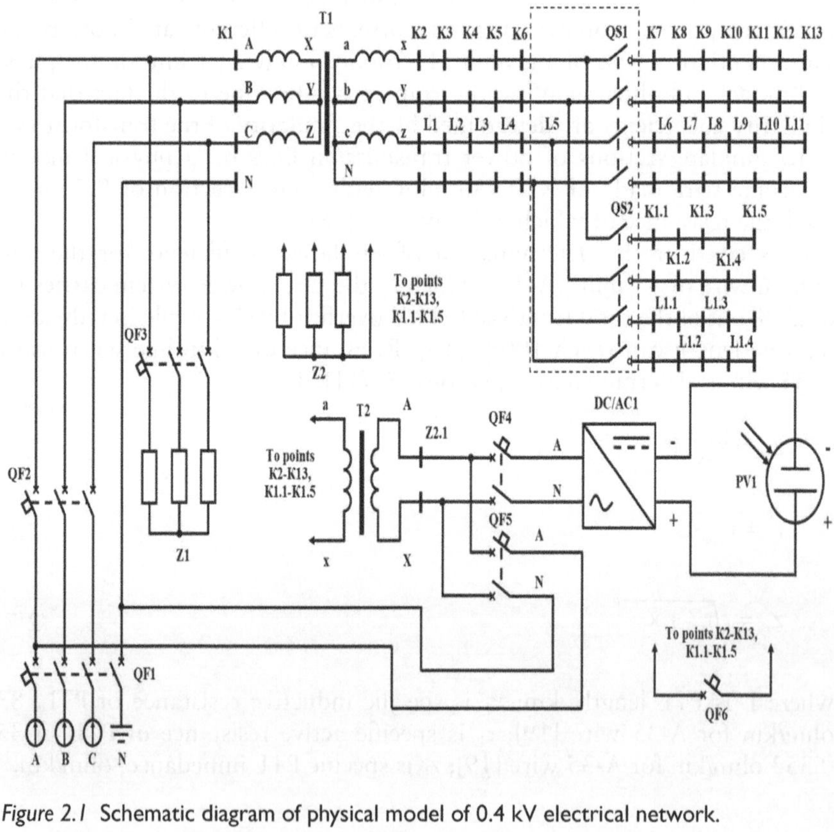

Figure 2.1 Schematic diagram of physical model of 0.4 kV electrical network.

circuit breaker; QS1, QS2 are three-pole load switches; Z1 is load on the higher voltage side of transformer T1; Z2 is mobile load for connection from the low-voltage side of transformer T1; T1 is three-phase transformer designed to power the physical model of electrical network; T2 is a single-phase transformer designed to connect a renewable power source and simulate reverse transformation; K1–K13, K1.1–K1.5 are terminals; L1–L11, L1.1–L1.4 are sections of LV PTLs; DC/AC1 is low power inverter; PV1 is solar panel.

To evaluate the reliability of the physical model, the experience of a stable single-phase short circuit on the low-voltage side of electrical network was implemented. The phase voltage on the model is 22 V which is electricity safe in accordance with national standards of many countries. So, In Russian Federation the safe AC voltage for humans is set to 42 V (the Rules for the Construction of Electrical Installations, a regulatory document in the field of electric power, approved by the Ministry of Energy of the Russian Federation). This paper presents a comparison of the obtained results of full-scale physical modeling of the specified mode with calculations made using the method of symmetrical components (specified in Russian standard GOST 28249-93 "Interstate standard. Short circuits in electrical installations. Calculation methods in AC electrical installations with voltage up to 1 kV"):

$$I^{1}_{SC} = \frac{\sqrt{3}\cdot U_{avg.LV}}{\sqrt{\left(2r_1 + r_0\right)^2 + \left(2x_1 + x_0\right)^2}}, \text{ kA}, \tag{2.9}$$

where $U_{avg.LV}$ is the average rated voltage of the network in which the short circuit occurred, V;r_1,x_1 are the total active and total inductive resistances of the direct sequence of short circuit, ohm;r_0,x_0 are the total active and total inductive resistances of the zero sequence of short circuit, ohm.

The calculation was carried out for the case of simulating a short circuit in a network containing ONAN-250 transformer and A-35 wire.

2.3 RESULTS AND DISCUSSION

The similarity coefficients of the low-power transformer TT-0.16-380/38 in the physical model and industrial prototypes of the transformer are determined by equations (2.1)–(2.4). The similarity coefficients obtained for transformers of different rated powers are presented in Table 2.3.

Similarity coefficients have been determined for wire A-35 using a transformer ONAN-250-10/0.4 kV Y/Y_n –0. The resistance per 1 km for an existing 0.4 kV PTL when using wire A-35 is:

$$x_{L(o)} = 0.352\cdot 1 = 0.352\ (\text{ohm});$$

Table 2.3 Similarity coefficients for6–10 kV power transformers of TM design types

	Transformers of ONAN design			
Power, kVA	*Voltage similarity coefficient, m_U, p.u.*	*Resistance similarity coefficient, m_Z, p.u.*	*Current similarity coefficient, m_I, p.u.*	*Power similarity coefficient, m_S, p.u.*
63	0.1	5.18	0.0193	0.00193
100	0.1	8.25	0.0121	0.00121
160	0.1	13.59	0.0074	0.00074
250	0.1	21.44	0.0047	0.00047

$$r_{L(o)} = 0.83 \cdot 1 = 0.83 \text{ (ohm)};$$

$$z_{L(o)} = \sqrt{0.83^2 + 0.352^2} = 0.902 \text{ (ohm)}.$$

Using the obtained similarity coefficients presented in Table 2.3, the resistance of 1 km of A-35 wire, $z_{L(M)}$, ohm, was determined for the physical model when using the ONAN-250 transformer power source as a prototype:

$$z_{L(M)} = z_{L(o)}^{A-35} \cdot k_z^{ONAN-250} \tag{2.10}$$

$$z_{L(M)} = 0.902 \cdot 21.44 = 19.34$$

The length of the operating line is 1 km on the model using nichrome wire with a cross section of 0.5 mm²with a resistance z_{X20H80} = 5.6 ohm/m, L_M, cm:

$$L_M = \frac{z_{L(M)} \cdot 100}{z_{X20H80}} \tag{2.11}$$

$$L_M = \frac{19.34 \cdot 100}{5.6} \approx 345 \text{ (cm)}$$

Figure 2.2 presents a dependence showing the length of nichrome wire with a cross section of 0.5 mm²with a resistance of 5.6 ohm/m, which is necessary to simulate 1 km of A-35 wires in terms of rated powers of transformers (TM design) from 25 to 630 kVA.

For example, 133 cm of nichrome wire with a cross section of 0.5 mm² is required to simulate 1 km of A-35 type wire when using a ONAN-100

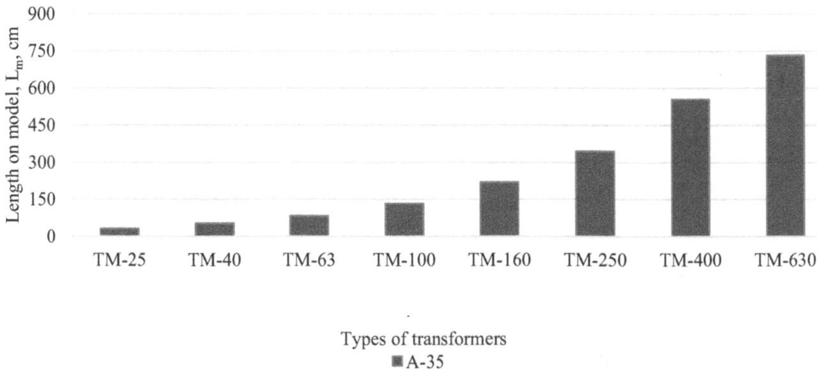

Figure 2.2 Dependence of the length of nichrome wire with a cross section of 0.5 mm^2 with a resistance of 5.6 ohm/m, which is necessary to simulate 1 km of A-35 wires in terms of rated powers of ONAN transformers from 25 to 630 kVA.

transformer as a prototype. At the same time, in a physical model, a length of 1 km of A-35 type wire requires a 556 cm length when using a ONAN-400 transformer as the power source.

Similar provisions are valid for other sections of bare, uninsulated wires, as well as aerial bundled cables, which are used on 0.4 kV overhead power lines.

In turn, load modeling is carried out in a similar way based on their resistance at an industrial voltage of 0.4 kV and the voltage adopted in the physical model. At the same time, a wide range of loads can be simulated on the physical model, which can be incandescent lamps, LED lamps, active-inductive loads, and loads of other types. The appearance of the developed physical model is presented in Figure 2.3.

To evaluate the reliability of the developed physical model, a single-phase short-circuit experiment was implemented. The short circuits were carried out sequentially at terminals K3–K13 of the physical model. Measurements were carried out with a certified digital multimeter OmixP99-MZ-3-0.5-RS 485. The measurement results are presented in Table 2.4.

As a result of the conducted measurements (Table 2.4), single-phase short-circuit currents were determined when creating a stable emergency mode at different points of the physical model—terminals K3–K13. Only in one of the experiments, when there was a short circuit at the K3 terminals, the single-phase short circuit current exceeded the rated phase current of the low-voltage winding of the transformer, and its load was 110%. To prevent failure of the transformer, a single-phase short circuit was not carried out at the transformer output at K2 terminals.

Figure 2.3 Appearance of the physical model of 0.4 kV electrical network.

Table 2.4 Results of measurements of single-phase short-circuit currents

No.	*Current of single-phase SC, A ($I_{SC}1$)*	*Full power of single-phase SC, VA ($S_{SC}1$)*	*SC location on the model*	*Resistance Z to SC location, Z_m, ohm*	*Transformer load at a given current, β, %*
1	No experiment was carried out to prevent transformer failure		K2	0	-
2	2.681	56	K3	3.6	110
3	2.08	44	K4	5	86
4	1.7	36	K5	6.4	70
5	1.443	31	K6	7.8	59
6	1.248	27	K7	9.2	51
7	1.1	24	K8	10.6	45
8	0.976	21	K9	12	40
9	0.87	19	K10	13.4	36
10	0.795	17	K11	14.8	33
11	0.731	16	K12	16.2	30
12	0.591	13	K13	19.8	24

Table 2.5 Results of recalculation of model parameters to an operating line when using a ONAN-250 transformer and A-35 wire

No.	*Current of single-phase SC, A ($I_{SC}1$)*	*Full power of single-phase SC, VA ($S_{SC}1$)*	*SC location on the model*	*Resistance Z to SC location on 0.4 kV PTL, Z_φ, ohm*	*Distance at SC on 0.4 kV PTL, L, m*
1	No experiment was carried out to prevent transformer failure (on model)		K2	0	0
2	570	119	K3	0.168	186
3	443	94	K4	0.233	258
4	362	77	K5	0.298	331
5	307	66	K6	0.363	402
6	266	57	K7	0.429	476
7	234	51	K8	0.494	548
8	208	45	K9	0.56	621
9	185	40	K10	0.625	693
10	169	36	K11	0.69	765
11	156	34	K12	0.756	838
12	126	28	K13	0.923	1023

The results of recalculation of single-phase short-circuit currents and short-circuit powers of the physical model to the operating line by equations (2.5) and (2.6), as well as resistances and distances in the operating line by equations (2.12) and (2.13), based on the data in Table 2.4, are presented in Table 2.5.

It should be noted that the results presented in Table 2.4 are universal and can be recalculated for any other type of transformer and wire. The results of full-scale modeling are comparable with the calculated results obtained by formula (2.9) using the method of symmetrical components using (GOST 28249-93). The calculation results and recalculated values measured on the model are presented in Figure 2.4.

It was found that the single-phase short-circuit current at a distance of 186 m from the 10/0.4 kV transformer is 570 A measured on the physical model (at point K3) while it equals to 549 A when calculating using the method of symmetrical components (by Russian standard GOST 28249-93), i.e., the error is 3.6%. In turn, at the most distant point at a distance of 1023 m, the results obtained using GOST and modeling methods were 118 and 126 A, respectively, i.e., error 6.3%.

Based on the research results presented in Figure 2.1, it was found that the single-phase short-circuit current at a distance of 186 m from the 10/0.4 kV transformer is 570 A measured on the physical model (at point K3) while it equals to 549 A when calculating using the method of symmetrical components (GOST 28249-93), i.e., the error is 3.6%. In turn, at the most distant point at a distance of 1023 m, the results obtained using GOST

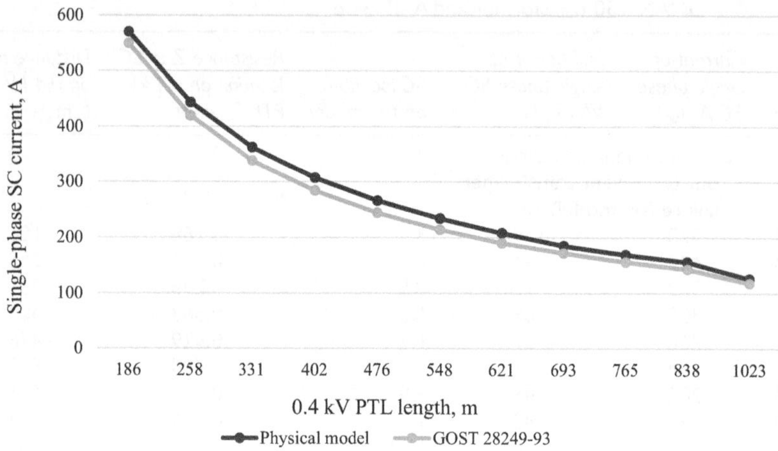

Figure 2.4 Single-phase SC currents determined by a developed physical model and calculated using the method of symmetrical components (GOST 28249-93), for the ONAN-250 transformer and A-35 wire.

and modeling methods were 118 and 126 A, respectively, i.e., error 6.3%. Thus, we can conclude that the results obtained from full-scale modeling on the developed stand have an error in the range of 3.6%–6.3% and the developed model is characterized by a fairly high degree of reliability.

2.4 CONCLUSIONS

Physical modeling of electrical networks makes it possible to study normal and emergency operating modes, the application of distributed generation sources in the network including renewable energy sources, and justifying the reliability of various methods of mathematical and computer modeling.

A physical model of 0.4 kV electrical network has been developed, which allows simulating real 0.4 kV electrical networks with different types and capacities of transformers and wires, sectionalizing, and redundancy means as well as generation sources. Using similarity theory, the similarity coefficients of the transformer ONAN-250-10/0.4 kV Y/Yn-0used in the model were determined in relation to ONAN power transformers of different powers. The lengths of nichrome wire made of X20N80 alloy with a crosssection of 0.5 mm^2 adopted in the model were also determined to simulate 1 km of 0.4 kV PTL made with different wires. The physical model of the 0.4 kV electrical network is electrically safe, as it has a combination of phase and phase-to-phase voltages of 22/38 V.

The experience of a stable single-phase short circuit at 11 different points was implemented on a physical model. The obtained current values were recalculated using similarity coefficients for a real 0.4 kV PTL. Comparison of the obtained values with ones calculated by the method of symmetrical components revealed that the relative error of the obtained results ranges from 3.6% to 6.3%, thereby indicating the reliability of the data obtained on the developed physical model.

REFERENCES

1. Folarin D.A. Modeling and simulation of faults in distribution network system using MATLAB/Simulink, *IOSR Journal of Electrical and Electronics Engineering*, Vol. 12, no. 3, pp. 43–51, 2018.
2. Singh, N., Tamrakar, S., Mewada, A., &Gupta, S.K. *Artificial Intelligence Techniques in Power Systems Operations and Analysis* (1st ed.), 2023, pp. 1–234. Auerbach Publications. https://doi.org/10.1201/978100330 1820
3. Martínez A., Navarro I., Quispe E., Donolo P., Santos V. MATLAB/Simulink modeling of electric motors operating with harmonics and unbalance, *International Journal of Electrical and Computer Engineering*, Vol. 12, pp. 4640–4648, 2022. doi:10.11591/ijece.v12i5.
4. Hussein A.S., Hawas M.N. Power quality analysis based on simulation and MATLAB/Simulink, *Indonesian Journal of Electrical Engineering and Computer Science*, Vol. 16, no. 3, pp. 1144–1153, December 2019. doi:10.11591/ijeecs.v16.i3.
5. Mlotok A.V., Ershov A.M., Valeev R.G., Sidorov A.I. Experimental electrical network with a voltage of 380 V, *Bulletin of the Engineering School of the Far Eastern Federal University*, Vol. 2, no. 19, pp. 96–107, 2014.
6. Valeev R.G., Mlotok A.V., Ershov A.M., Sidorov A.I. Modeling of a 380 V electrical network with overhead lines in the MATLAB-Simulink software environment. News of higher educational institutions, *Energy Problems*, Vol. 9, no.10. pp. 116–128,2013.
7. Ershov A.M., Khlopova A.V. Physical model of an electrical network with a voltage of 10/0.38 kV, *Electrical Safety*, Vol. 2, pp. 13–21,2016.
8. Vinogradov A.V., Lansberg A.A. Experience of measurements during a single-phase short circuit in a 0.4 kV electrical network, *Agricultural Technology and Energy Supply*, Vol. 3, no. 36, pp. 5–15, 2022.
9. Lansberg A.A. Justification of the parameters of a computer model of a rural electrical network 0.4 kV in MATLAB SIMULINK, *Agricultural Technology and Energy Supply*, Vol.3, no. 40, pp. 36–47,2023.
10. *Relay Protection and Automation Project. Relay Protection Laboratory.* https://pro-rza.ru/laboratoriya-rza/ (accessed: August 22, 2023).
11. Balabin A.A., Vinogradov A.V., Lansberg A.A. Analysis of work and recommendations for improving the electrical energy storage system installed in a 0.4 kV rural electrical network, *Agroengineering*, Vol. 24, no. 1, pp. 72–79, 2022.

12. Rules for the installation of electrical installations [Text]/Federal Environmental and Technological Service. and atomic control. 7th ed. Saint Petersburg: DEAN, 2013. - 701 p., [2] L. maps : ill., table.; 21 cm. (Occupational safety in Russia).; ISBN 978-5-93630-923-6 (in translation).
13. Vinogradov A.V., Lansberg A.A., Sorokin N.S. Characteristics of electric grid companies by the number and length of power transmission lines, power of substations, In: *Electrical Technologies and Electrical Equipment in the Agro-Industrial Complex*, Vol. 69, no. 2(47), pp. 31–41, 2022.
14. Venikov V.A. *Theory of Similarity and Modeling (in Relation to Problems in the Electric Power Industry)*, Moscow, Russia, 1976, High School.
15. Ershov A.M., Sidorov A.I., Valeev R.G. *Protection of Electrical Networks with a Voltage of 380 V During Single-Phase Short Circuits: Monograph*, Vologda, Russia, 2021, Infra-Engineering, 232pp.
16. Vinogradova A.V., Lansberg A.A., Vinogradov A.V. *Energy System of the Oryol Region: Review of Statistical Information*, Orel, Russia, 2023, Kartush, 360pp.
17. *New Technologies. Power Transformers, Substations, Electrical Equipment, etc.* https://transelektro.nt-rt.ru/?ysclid=lfwdlojypo86783165 (accessed: March 31, 2023).
18. *Nichrome. Nichrome Wire, Coil 10 m. Chipdip*, www.chipdip.ru/product/nihrom-0.4?ysclid=lftnwbtl9u423173563 (accessed: April 26, 2023).
19. Makarov, E.F. *Handbook on Electrical Networks of 0.4–35 kV and 110–1150 kV*, Edited by I.T. Goryunova and A.A. Lyubimova, Moscow, Russia, 2003, Papyrus PRO.

Chapter 3

Hybrid BPSO–GTO algorithm to control stochastic energy management system under diverse load conditions

Neha Smitha Lakra and Baidyanath Bag

3.1 INTRODUCTION

The scarcity of fossil fuel reserves, environmental concerns, and the need to meet power demand have compelled distribution system operators (DSOs) to deploy energy management techniques. These strategies aim to uphold a flat voltage profile across each bus, maximize efficiency and cost savings, and mitigate network loss. In recent years, conservation voltage reduction (CVR) has emerged as a proven method for lowering peak load and power losses. CVR endeavors to maximize long-term energy savings while reducing voltage within American National Standards Institute (ANSI) standards [1]. Various utilities across the U.S. have deployed this method to reduce power demand and losses [2]. An overview of CVR's performance, benefits, and methods has been elaborated in reference [3]. While CVR inherently operates the distribution system within a lower voltage limit, it may partially mitigate over-voltage issues. However, CVR alone can cause voltage instability problems and increase or decrease line losses depending on a distribution system's parameters and properties. Therefore, to address these challenges, integrating CVR with network reconfiguration (NR) and distributed generation (DG) presents a viable solution for enhancing energy efficiency. However, unpredictable load behavior introduced complexity to the proposed energy management scheme. NR involves modifying the network topology by opening and closing switches while maintaining a radial structure [4]. It provides numerous benefits, such as reducing losses, enhancing reliability, restoring service, managing load congestion, alleviating overloads, and improving voltage profiles. Various algorithms have discussed NR with different objectives, such as fractal search algorithm (FSA) [5], modified particle swarm optimization (MPSO) [6], firefly algorithm (FA) [7], salp swarm algorithm (SSA) [8], grey wolf optimization (GWO) [9], and more. Moreover, non-conventional sources are projected to account for approximately 18% of global power generation by 2040 [10]. DG serves as a greener alternative to fossil fuel and clean energy. It provides a multitude of benefits, including enhanced voltage

DOI: 10.1201/9781003516156-3

profiles [11], reduced power losses [12], and alleviation of congestion and deferred investments [13]. Therefore, DG integration has risen drastically in recent years. However, inaccurate placement and sizing of DG can adversely impact the system's performance. In the existing literature, various optimization algorithms with different objectives have been applied for DG placement. These methods include the stud krill herd algorithm (SKHA) [14], water cycle algorithm (WCA) [15], PSO [16], ant colony optimization [11], bat algorithm [17], and so on. In reference [18], authors introduced a plant growth simulation (PGS) algorithm in conjunction with PSO to determine the best switch configurations for NR and optimal placement of DG. Similarly, in reference [19], the authors addressed the combined operation of NR and DG by utilizing an improved SSA.

Nonetheless, studying the systems' behavior with various load types is imperative for optimal effective energy efficiency. The distribution system includes various loads, each with unique characteristics, mainly including industrial (IN), commercial (CO), and residential (RE) loads. This section provides an overview of various types of loads considered in energy-efficient techniques. Authors of [20] has deployed a CVR method to minimize power usage and peak demand. This chapter reported that a 1% reduction in voltage produced energy savings of upto 0.4%. Similarly, CVR is tested on an actual 8500-bus system with an RE load to curtail peak demand [21]. In reference [22], the Volt/VAr control (VVC) scheme was verified on the IEEE-33 bus to reduce power consumption and losses while accounting for constant impedance (Z) load. Similarly, in reference [23], authors applied the VVC method to the same system but with a ZIP load type. The authors in reference [24] presented a VVC method to maximize energy savings while incorporating a ZIP load model. The combined scheme of DG and NR was presented in reference [25] to reduce line losses by utilizing PQ and time-varying load for demand representation. The method integrating CVR with DG and NR has been presented in reference [13] to minimize the energy demand of the network while considering the ZIP load model. In addition, reference [26] proposed the integrated method of capacitor placement and NR to decrease the total investment costs, using PQ and ZIP load to depict demand. Further, the integrated operation of VVC, NR, and DG has been conducted in reference [27] to curtail power demand and losses while considering exponential load modeling (ELM) and ZIP loads. The coordinated scheme of on-load tap changer (OLTC) with NR and DG has been implemented to decrease the line losses [28], utilizing PQ and ZIP loads.

3.2 PROBLEM FORMULATION

This chapter proposes a methodology to coordinate the stochastic operation of the CVR, NR, and DG considering load type and loading levels. The main goal is to decrease power demand by reducing the voltage levels nearer to the ANSI lower limits.

3.2.1 Objective: Power demand reduction

The objective is to reduce the system's power demand, which is determined by summing the power consumption of different consumers and the losses in the connecting lines. It is delineated as follows:

$$P_{Demand} = \left\{ \sum_{i=1}^{N_{bus}} P_{cons,i} + \sum_{i=1}^{N_{bus}-1} P_{loss,i} \right\} \tag{3.1}$$

The power consumption and power losses are evaluated by equations (3.2) and (3.3), respectively.

$$P_{cons,i} = \mu\, P_0 \left(\frac{U_i}{U_0} \right)^{k_i^P} \tag{3.2}$$

$$P_{loss,i} = \left(U_i - U_j \right)^2 \times \frac{r_{ij}}{z_{ij}^2};\ i \neq j \tag{3.3}$$

where P_{Demand} is overall power demand; N_{bus} is the number of buses; i, j are indices for bus number; $P_{cons,i}$ is power consumption at ith bus; $P_{loss,i}$ is active power loss at ith branch; μ is loading levels; P_0 is the nominal value of power consumption at ith bus; U_i, U_j is voltage magnitude at ith and jth bus, respectively; U_0 is the nominal rating of voltage at ith bus; k_i^P is the exponent value of real power consumption; r_{ij}, z_{ij} is resistance and impedance of the line connecting ith and jth bus, respectively.

3.2.2 Constraints

3.2.2.1 Power flow constraints

To ensure steady operation of an electrical system, the power flow equation must be satisfied. It can be stated as follows:

$$P_{Sub} + \sum_{i=1}^{N_{DG}} P_{DG,i} = \sum_{i=1}^{N_{bus}} P_{cons,i} + \sum_{i=1}^{N_{bus}-1} P_{loss,i} \tag{3.4}$$

$$Q_{Sub} = \sum_{i=1}^{N_{bus}} Q_{cons,i} + \sum_{i=1}^{N_{bus}-1} Q_{loss,i} \tag{3.5}$$

where P_{Sub}, Q_{Sub} and $P_{cons,i}, Q_{cons,i}$ are real and reactive power generation and consumption at *i*th bus, respectively; N_{DG} is the number of buses for installing DG bus; $Q_{loss,i}$ is reactive power loss at *i*th branch.

3.2.2.2 Bus voltage constraints

The magnitude of bus voltage should lie within the permissible limits as follows:

$$U_i^{LB} \leq U_i \leq U_i^{UB} \tag{3.6}$$

where U_i^{LB}, U_i^{UB} is the lower and upper limit at *i*th bus.

3.2.2.3 OLTC transformer tap constraints

It is used in the system to regulate voltage levels of the distribution system. The mathematical representation of OLTC is expressed using equation (3.7):

$$U_{tap} = 1 + tap \frac{\Delta U_{step}}{100} \tag{3.7}$$

where $tap \in \{-8,\ldots,-1,0,1,\ldots,+8\}$, U_{tap} is voltage magnitude at the secondary side of OLTC; tap is the OLTC taps; ΔU_{step} is variation in voltage value with a change in tap position.

3.2.2.4 DG constraints

The number of candidate buses for installing DG must be less than the total bus and is expressed using equation (3.8). Furthermore, power supplied by DG must be within permissible bounds; DG should contribute power less than overall demand, as presented in equations (3.9) and (3.10), respectively:

$$N_{DG} \leq N_{bus} \tag{3.8}$$

$$P_{DG,i}^{LB} \leq P_{DG,i} \leq P_{DG,i}^{UB};\ i \neq 1 \tag{3.9}$$

$$\sum_{i=1}^{N_{DG}} P_{DG,i} \leq \sum_{i=1}^{N_{bus}} P_{cons,i} + \sum_{i=1}^{N_{bus}-1} P_{loss,i} \tag{3.10}$$

3.2.2.5 *Radiality constraints*

The rank of the branch-bus incidence matrix (BBIM) should be equal to the number of branches to maintain the radiality of the system as given in equation (3.11). Each element of the *BBIM*) order of $N_{bus} \times N_{bus}$ is formed using equation (3.12):

$$\text{rank}\left(BBIM\right) = nbr \tag{3.11}$$

$$BBIM = \begin{cases} 1, & \textit{if node i is connected to node j} \\ 0, & \textit{otherwise} \end{cases} \tag{3.12}$$

3.2.2.6 *Load models and loading levels*

In reality, the distribution system comprises a wide range of loads, such as IN, CO, RE, and mixed load. They significantly impact the distribution system's effectiveness. Furthermore, changes in loading levels can impact system performance. Therefore, various static load models and different loading levels have been considered to depict the actual load accurately. The mathematical expression for load consumption is as follows:

$$P_{cons,i} = \mu P_0 \left(\frac{U_i}{U_0}\right)^{k_i^p}; \ \mu \in \{0.5, 1, 1.6\} \tag{3.13}$$

$$Q_{cons,i} = \mu Q_0 \left(\frac{U_i}{U_0}\right)^{k_i^q}; \ \mu \in \{0.5, 1, 1.6\} \tag{3.14}$$

3.3 PROPOSED APPROACH

This chapter aims to improve the efficiency of the distribution network's power demand through the implementation of energy-saving techniques. The approach integrates CVR, NR, and DG. It formulates a mixed integer non-linear problem (MINLP) that incorporates OLTC tap adjustments, switch configurations, and the placement and capacity determination of DG units. A hybrid BPSO–GTO algorithm is utilized to tackle this: BPSO

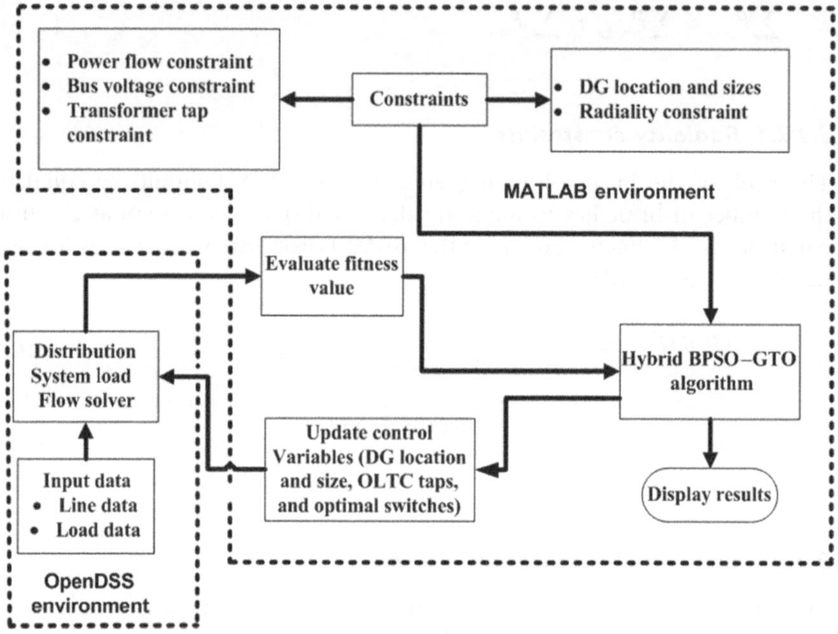

Figure 3.1 Generalized block diagram of energy-efficient approach.

optimizes switch settings, while GTO refines OLTC tap positions and DG specifications. The generalized block diagram of the proposed methodology is visualized in Figure 3.1.

3.3.1 BPSO algorithm for network reconfiguration

BPSO is a modification of the PSO algorithm where the solution vector is represented in the binary format. Each swarm is characterized by a position and velocity within d-search space. The swarm's personal best is represented as $P_{besti} = \left(P_{besti1}, P_{besti2}, \ldots, P_{bestid}\right)$, and G_{besti} represents best of P_{besti}. At every iteration, a swarm adjusts its velocity using equation (3.15):

$$V_i^{t+1} = wV_i^t + C_1 r_1 \left(P_{best,i}^t - P_i^t\right) + C_2 r_2 \left(G_{best,i}^t - P_i^t\right) \tag{3.15}$$

where t and $t+1$ denote current and next iteration, C_1 and C_2 denote acceleration factors, r_1 and r_2 are random variables restricted between 0 and 1and w is the inertia weight updated according to equation (3.16):

$$w = w_{max} - \frac{w_{max} - w_{min}}{t_{max}} \times t \tag{3.16}$$

where t_{max} is the maximum iteration, w_{max} and w_{min} are upper and lower weight values, respectively. In BPSO, the particles' velocities and positions are updated according to equations (3.17) and (3.18), respectively:

$$V_i^{t+1} = \frac{1}{1+e^{-V_i^t}} \tag{3.17}$$

$$P_i^{t+1} = \begin{cases} 1 & if\ z_i \le \frac{1}{1+e^{-V_i^t}} \\ 0 & otherwise \end{cases} \tag{3.18}$$

where z_i is auniform random number between 0 and 1.

3.3.2 GTO algorithm for energy management

GTO is a novel metaheuristic algorithm derived from the social behavior of gorillas [29]. The algorithm consists of two phases: exploration and exploitation. Five distinct operators simulate gorilla behavior. The exploration stage consists of three operators: moving toward an unknown place, migrating toward other gorillas, and relocating to a familiar place. In contrast, the exploitation stage consists of trailing the silverback and competing for adult females.

3.3.2.1 *Exploration*

In this section, three exploration mechanisms are presented and denoted by equation (3.19). In the equation, *p* denotes parameters ranging from 0 to 1, which determines the probability of selecting a move toward an unknown place. However, if $rand \ge 0.5$, migration toward other gorillas is chosen. However, if $rand < 0.5$, moving to a known place is chosen:

$$GX(t+1) = \begin{cases} (UL - LL) \times r_1 + LL, & rand < p \\ (r_2 - C) \times X_r(t) + L \times H, & rand \ge 0.5 \\ X(i) - L \times \big(L \times (X(t) - GX_r(t)) + r_3 \times (X(t) - GX_r(t))\big), & rand < 0.5 \end{cases} \tag{3.19}$$

where $X(t)$ and $GX(t+1)$ denote the current and subsequent positions of the gorilla. Moreover, r_1, r_2, r_3 and *rand* are random variables between

0 and 1. UL and LL denote the upper and lowerbounds of the variables, respectively. X_r and GX_r denote randomly selected gorillas from the whole group. The values of $C, L,$ and H are estimated based on the following equations:

$$C = F \times \left(1 - \frac{iter}{t_{max}}\right) \tag{3.20}$$

F is calculated using equation (3.21):

$$F = \cos(2 \times r_4) + 1 \tag{3.21}$$

where r_4 is a random number restricted between −1 and 1. L can be evaluated using equation (3.22):

$$L = C \times l \tag{3.22}$$

where l is a random number in the range (0–1). Additionally, H is computed using equation (3.23). Again, Z is estimated using equation (3.24), where Z is a random value within a range of $-C$ to C.:

$$H = Z \times X(t) \tag{3.23}$$

$$Z = [-C, C] \tag{3.24}$$

After the exploration mechanism is completed, a group operation is performed to evaluate the cost of all GX solutions. If the $GX(t) < X(t)$, then replace $X(t)$ with $GX(t)$. Thus, the finest solution achieved in this stage is also recognized as a silverback.

3.3.2.2 Exploitation

This section describes the silverback's role and the competition for adult females during the exploitation stage. The silverback, as the primary leader, guides the group and makes all decisions. The young gorillas in the group follow the silverback to locate their prey. That is to say, if $C \geq W$, the tracking of the silverback will be executed. Equation (3.25) is used to depict this process, as follows:

$$GX(t+1) = L \times M \times \left(X(t) - X_{silverback}\right) + X(t) \tag{3.25}$$

$$M = \left(\left| \frac{1}{N} \sum_{i=1}^{N} GX_i(t) \right|^g \right)^{\frac{1}{g}} \tag{3.26}$$

$$g = 2^L \tag{3.27}$$

where $X(t)$ and $X_{silverback}$ denote the vector position of the gorilla and silverback, respectively. Moreover, L is evaluated by equation (3.22) and M by equation (3.26), where $GX_i(t)$ denotes each gorilla's position tth iteration; N denotes a total number of gorillas; g is evaluated using equation (3.27).

However, if $C < W$, the fight among young gorillas for adult females is executed. Equation (3.28) is used to depict this behavior:

$$GX(i) = X_{silverback} - (X_{silverback} \times Q - X(t) \times Q) \times A \tag{3.28}$$

$$Q = 2 \times r_5 - 1 \tag{3.29}$$

$$A = \beta \times E \tag{3.30}$$

$$E = \begin{cases} N_1, & rand \geq 0.5 \\ N_2, & rand < 0.5 \end{cases} \tag{3.31}$$

where Q denotes impact force, r_5 denotes a random valuewithin (0–1). Equation (3.30) evaluates the degree of violence in conflict, with β being a parameter defined before optimization. E represents violence's impact on the solution dimensions. On completing the exploitation stage, a group operation estimates the cost of all GX solutions. If the $GX(t) < X(t)$, then replace $X(t)$ with $GX(t)$. Thus, the finest solution obtained in this stage is recognized as a silverback.

3.4 CASE STUDY AND SIMULATION RESULTS

The usefulness of the proposed method has been evaluated using the IEEE-69 bus network, which includes 68 sectionalizing switches and 5 tie-switches, as depicted in Figure 3.2. The data information of this network is derived from reference [30]. The proposed approach is implemented using the MATLAB COM interface of OpenDSS [31]. To address the NR and DG placement problems, the methodology applies BPSO and GTO techniques, respectively.

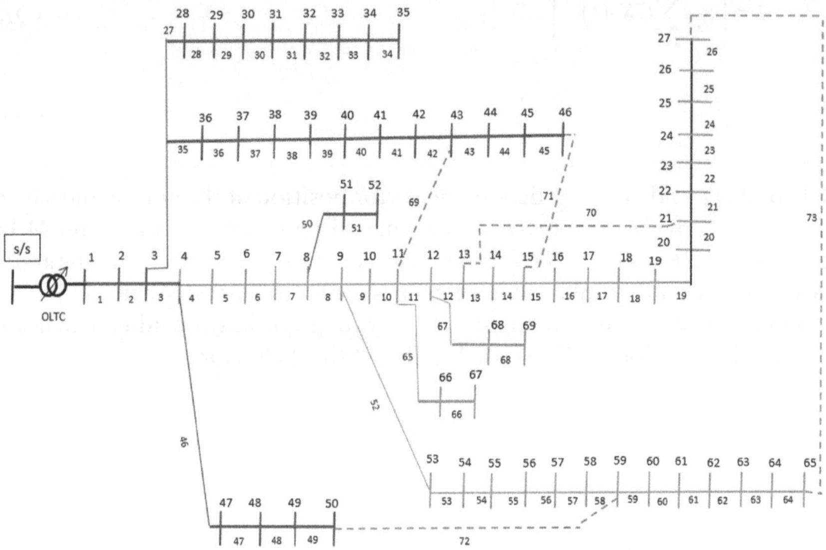

Figure 3.2 Network structure of test system.

Table 3.1 Voltage coefficient of different load types

Load type	k_i^p	k_i^q
PQ	0	0
IN	0.18	6.00
CO	1.50	3.15
RE	1.04	4.19

Table 3.1 depicts the values of the voltage coefficient of different loads. Table 3.2 presents the parameters of the hybrid BPSO–GTO algorithm. The investigation of the proposed approach includes four different cases which are presented below:

1. Base case
2. OLTC with NR
3. OLTC with DG
4. OLTC with NR and DG

The impact of different load types and loading levels is explored for each case. The simulated results for IN, CO, RE, and ELM are presented in Table 3.3.

Table 3.2 Parameters of a hybrid BPSO–GTO algorithm

Parameters	*BPSO*	*GTO*
Population size, *Pop*	20	20
Maximum iteration, t_{max}	50	50
C_1, C_2	2,2	–
w_{max}, w_{min}	0.9,0.4	–
β	–	3
p	–	0.03
Optimal switches	5	-
OLTC taps	16	16
Location of DG	-	3

Table 3.3 Results of different loads

Cases		*Load type*			
		IN	*CO*	*RE*	*Mixed*
Base case	Optimal switches	69,70,71,72,73	69,70,71,72,73	69,70,71,72,73	69,70,71,72,73
	P_{loss} (kW)	180.93	179.10	179.20	179.19
	P_{cons} (kW)	3978.4	3949.9	3957.9	3971.4
	P_{Demand} (kW)	4159.33	4129	4137.1	4150.59
	U^{LB}(p.u.)	0.9555	0.9569	0.9565	0.9565
OLTC with NR	Optimal switches	14,55,61,69,70	14,55,61,69,70	14,55,61,69,70	14,55,61,69,70
	P_{Loss} (kW)	85.03	83.80	83.94	83.94
	P_{cons} (kW)	3869.6	3742.8	3786.3	3780.5
	P_{Demand} (kW)	3954.63	3826.6	3870.24	3864.44
	U^{LB} (p. u.)	0.9540	0.9540	0.9541	0.9541
OLTC with DG	Optimal switches	69,70,71,72,73	69,70,71,72,73	69,70,71,72,73	69,70,71,72,73
	DG capacity (kW)	61(1520) 11(380) 18(380)	61(1518) 11(190) 18(380)	61(1483.2) 11(381.5) 18(381.9)	61(1483) 11(380) 18(380)
	Total DG capacity (kW)	2281	2088	2246.6	2243
	P_{Loss} (kW)	44.82	56.51	56.39	51.40
	P_{cons} (kW)	3775.9	3587.39	3652.91	3640.7
	P_{Demand} (kW)	3820.72	3643.9	3709.3	3692.1
	U^{LB} (p.u.)	0.9514	0.9517	0.9513	0.9512
OLTC with NR and DG	Optimal switches	14,55,61,69,70	14,55,61,69,70	14,55,61,69,70	14,55,61,69,70
	DG capacity (kW)	61(1342.7) 11(380.3) 64(474.3)	61(1294.6) 11(524.3) 64(449)	61(1288.2) 11(479.6) 64(466.4)	61(1360.9) 11(414.7) 64(441.9)
	Total DG capacity (kW)	2197.2	2267.9	2234.2	2217.6
	P_{Loss} (kW)	24.58	28.96	27.11	27.38
	P_{cons}(kW)	3777.42	3598.84	3660.39	3647.22
	P_{Demand} (kW)	3802	3627.8	3687.5	3674.6
	U^{LB} (p.u.)	0.9584	0.9572	0.9573	0.9586

3.4.1 Comparative analysis of various voltage-dependent loads

This section explicitly discusses the effect of the proposed method on the voltage behavior, power losses, power consumption, and power demand of the network across different load models. Moreover, the correlation between power consumption and voltage levels has been analyzed for the above-discussed load models.

3.4.1.1 Voltage profile

The impact of different operational cases on the voltage profile of various loads is examined, as presented in Figure 3.3. Each case demonstrates distinct characteristics and influences specific load types, resulting in diverse voltage profiles. Notably, case 4 achieves the highest voltage reduction within a lower permissible limit for all load types.

3.4.1.2 Reduction in power losses

Based on Figure3.4, case 4 consistently achieves the highest reduction in power loss across all load types. However, different load types show varying

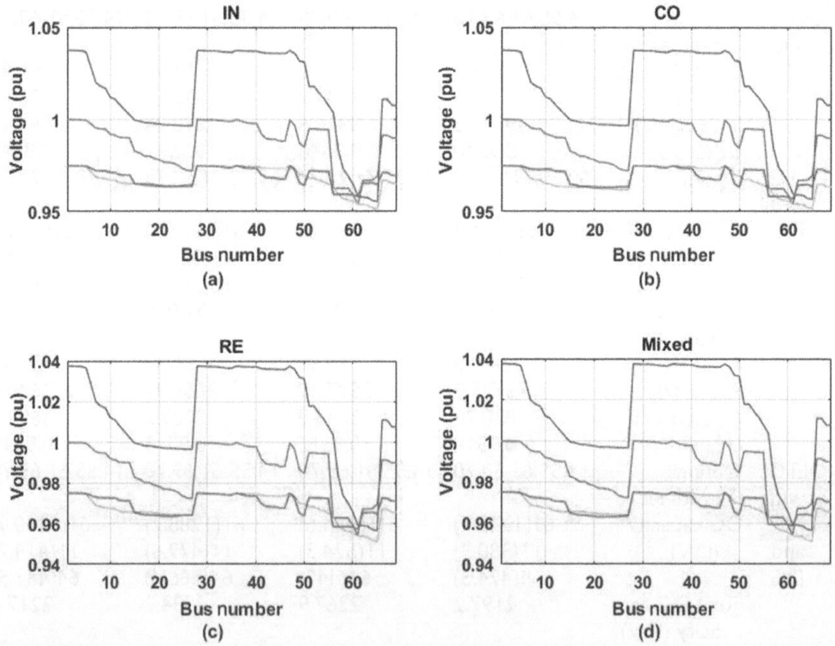

Figure 3.3 Voltage profile of different voltage-dependent loads.

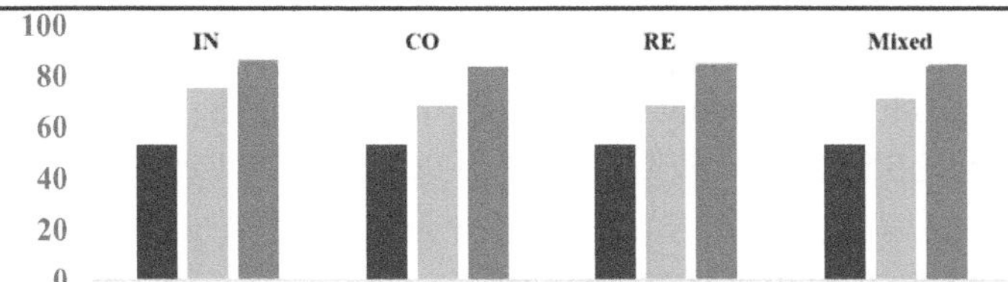

	P_{loss} reduction (%)	P_{loss} reduction (%)	P_{loss} reduction (%)	P_{loss} reduction (%)
C-2	53	53.21	53.16	53.16
C-3	75.23	68.45	68.53	71.32
C-4	86.41	83.83	84.87	84.72

Figure 3.4 Power loss reductions of different voltage-dependent loads.

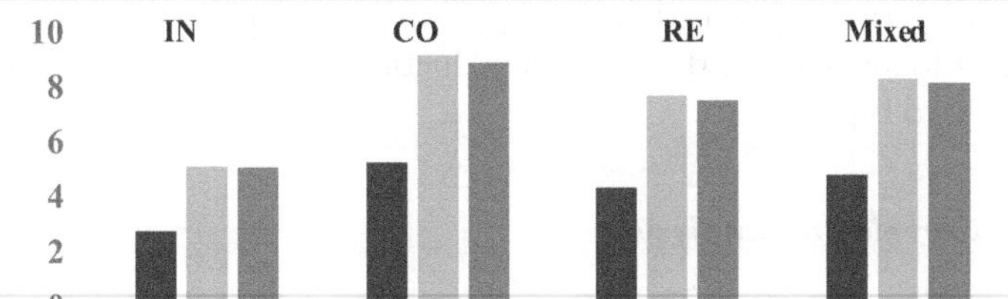

	P_{cons} reduction (%)	P_{cons} reduction (%)	P_{cons} reduction (%)	P_{cons} reduction (%)
C-2	2.73	5.24	4.33	4.8
C-3	5.09	9.17	7.7	8.32
C-4	5.05	8.88	7.51	8.16

Figure 3.5 Power consumption reductions of different voltage-dependent loads.

loss values due to their voltage exponent values in Table 3.1. In case 2, CO loads show the most significant power loss reductions, which is attributed to their sensitivity to voltage levels, resulting in decreased power consumption and reduced power loss. In cases 3 and 4, IN loads exhibit the maximum power loss reduction due to the integration of DG alone or combined, significantly raising the voltage profile and thus increasing consumption for all loads except IN loads, resulting in substantial power loss reductions for IN loads.

3.4.1.3 Reduction in power consumption

Based on the data presented in Figure 3.5, case 4 consistently yields the greatest reduction in power consumption across all load types. However, the specific load types exhibit different consumption values due to variances in their voltage exponent values, as specified in Table 3.1. For all cases, the CO loads stand out with the highest reduction in power consumption. This can be attributed to the CO loads having the lowest voltage profile among the different load types, resulting in a more significant decrease in power consumption.

3.4.1.4 Reduction in power demand

Based on Figure 3.6, case 4 significantly achieves the highest power demand reduction for all load types. The data presented in this figure also reveal that different load types exhibit varying power demand values due to variations in their voltage exponent values.

Power demand reduction, a key objective of this study, is formulated using equation (3.1), which is the summation of power consumption by different consumers and power losses of the connecting lines. Examining Figures 3.4 and 3.5, it is notable that for case 2, CO loads exhibit the maximum reduction in power loss and power consumption. Therefore, in all cases, CO loads achieve the greatest reduction in power demand compared to other load types.

3.4.1.5 Correlation analysis

In this section, Figure 3.7 presents a correlation heat map illustrating the relationship between power consumption and voltage for various loads. This analysis aims to assess the impact of voltage fluctuations on power consumption for these loads. For simplicity, the base case with no control measures has been chosen. The figure indicates that, among all loads, the CO load exhibits a stronger correlation between voltage and power

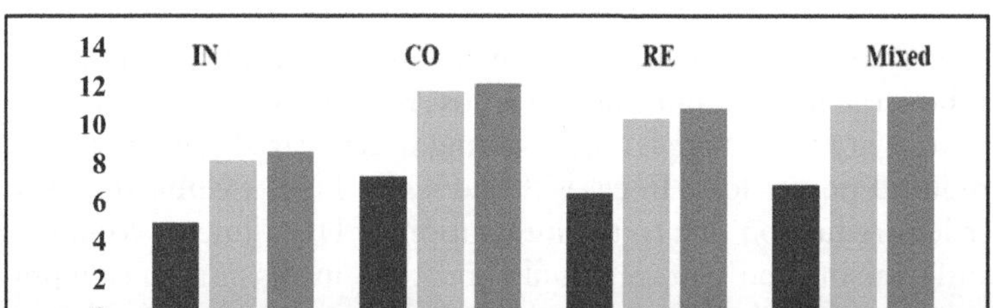

	Demand reduction (%)	Demand reduction (%)	Demand reduction (%)	Demand reduction (%)
■C-2	4.92	7.32	6.45	6.89
■C-3	8.14	11.75	10.34	11.05
■C-4	8.59	12.14	10.87	11.47

Figure 3.6 Demand reductions of different voltage-dependent loads.

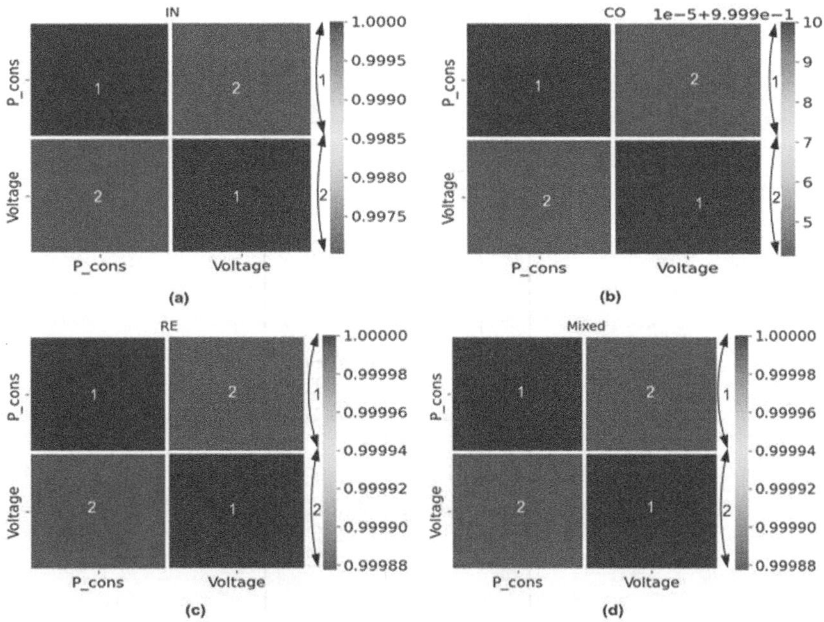

Figure 3.7 Correlation heat map of different load types.

consumption. Consequently, CO loads result in a more significant reduction in power demand compared to other loads.

3.4.2 Effect of loadings

The proposed approach (case 4) has been evaluated under normal loading conditions, with an analysis considering various load models. However, to fully understand its dynamic behavior, it is also essential to examine it under low and heavy loading conditions. Table 3.4 illustrates the impact of low and heavy loading conditions on the proposed approach across various load models.

Specifically, real and reactive demands are adjusted to 50% and 160% of the total connected loads during low and heavy loading conditions, respectively. The table reveals that the proposed scheme maintains superior performance despite fluctuations in loading conditions across all load models.

3.5 CONCLUSION

This chapter develops various load models to assess the impact of the coordinated approach of CVR, NR, and DG on the distribution system's

Table 3.4 Effect of loading on the proposed scheme across load models

	IN		*CO*		*RE*		*Mixed*	
	Loading		*Loading*		*Loading*		*Loading*	
Parameters	*Low*	*Heavy*	*Low*	*Heavy*	*Low*	*Heavy*	*Low*	*Heavy*
P_{loss} (kW)	7.03	89.43	7.9	95.29	7.40	93.36	7.42	93.43
P_{cons} (kW)	1885.67	6055.17	1776.12	5851.91	1813.32	5920.84	1804.68	5908.67
P_{demand} (kW)	1892.7	6144.6	1784.02	5947.2	1820.72	6014.2	1812.1	6002.1
U^{LB} (pu)	0.9502	0.9588	0.9501	0.9591	0.9504	0.9585	0.9503	0.9585

energy efficiency. The analysis includes IN, CO, RE, and mixed loads. The hybrid BPSO–GTO algorithm is utilized to determine the optimal positioning and sizing of OLTC taps, opened switches, and DG for each load model. The proposed approach is tested on the IEEE-69 bus system under different cases, including light and heavy loading conditions. The results demonstrate significant benefits such as reduced power demand, lower power losses, decreased load consumption, and improved voltage profiles. These benefits are sustained even under heavy loading levels. The study finds that different load models significantly affect OLTC tap settings, switch configurations, and DG, leading to varying power demands and voltage profiles. A correlation analysis shows that CO loads have a higher correlation value, leading to greater power consumption reduction across all loading levels. The proposed approach is particularly effective for CO loads, maximizing energy efficiency. These findings will guide system operators and decision-makers in selecting appropriate load models to optimize energy efficiency in distribution systems. Additionally, implementing this energy-efficient approach promotes sustainable and eco-friendly energy for consumers.

REFERENCES

[1] N. Singh, S. Tamrakar, A. Mewada, and S. K. Gupta, *Artificial Intelligence Techniques in Power Systems Operations and Analysis* (1st ed.), New York: Auerbach Publications, 2023, pp. 1–23. https://doi.org/10.1201/9781003301820

[2] Z. S. Hossein, S. Member, A. Khodaei, and S. Member, "Conservation Voltage Reduction and Volt-VAR Optimization: Measurement and Verification Benchmarking," *IEEE Access*, vol. 8, pp. 50755–50770,2020.

[3] N. Singh and Y. Kumar, "Optimization Techniques Used for Designing Economic Electrical Power Distribution," *Meta-heuristic Optimization Techniques: Applications in Engineering*, edited by Anuj Kumar, Sangeeta Pant, Mangey Ram, and Om Yadav, Berlin, Boston: De Gruyter, 2022, pp. 79–92. https://doi.org/10.1515/9783110716214-005

[4] H. Jia, X. Zhu, and W. Cao, "Distribution Network Reconfiguration Based on an Improved Arithmetic Optimiz ation Algorithm," *Energies*, vol. 17, no. 8, p. 1969, 2024.

[5] N. Singh, A. Nighoskar, and S. Ram Krishna, "Economic Load Dispatch with Valve Loading Effects Optimize by BBMO," *Int. J. Innov. Technol. Explor. Eng.*, vol. 8, no. 7, pp.104–107, 2019.

[6] S. Essallah and A. Khedher, "Optimization of Distribution System Operation by Network Reconfiguration and DG Integration Using MPSO Algorithm," *Renew. Energy Focus*, vol. 34, pp. 37–46, 2020.

[7] N. Singh, T. Chakrabarti, P. Chakrabarti, M. Margala, A. Gupta, S. P. Praveen, S. B. Krishnan, and B. Unhelkar,"Novel Heuristic Optimization Technique to Solve Economic Load Dispatch and Economic Emission Load Dispatch Problems," *Electronics*, vol.12, no.13, p. 2921,2023. https://doi.org/10.3390/electronics12132921

[8] K. S. Sambaiah and T. Jayabarathi, "Optimal Reconfiguration and Renewable Distributed Generation Allocation in Electric Distribution Systems," *Int. J. Ambient Energy*, vol. 42, no. 9, pp. 1018–1031, 2021.

[9] H. K. Pujari and M. Rudramoorthy, "Grey Wolf Optimisation Algorithm for Solving Distribution Network Reconfiguration Considering Distributed Generators Simultaneously," *Int. J. Sustain. Energy*, vol. 41, no. 11, pp. 2121–2149,2022.

[10] N. Singh and Y. Kumar, "Optimization Techniques Used for Designing Economic Electrical Power Distribution," *Meta-heuristic Optimization Techniques: Applications in Engineering*, edited by Anuj Kumar, Sangeeta Pant, Mangey Ram, and Om Yadav, Berlin, Boston: De Gruyter, 2022, pp. 79–92. https://doi.org/10.1515/9783110716214-005

[11] N. R. Godha, V. N. Bapat, and I. Korachagaon, "Ant Colony Optimization Technique for Integrating Renewable DG in Distribution System with Techno-economic Objectives," *Evol. Syst.*, vol. 13, no. 3, pp. 485–498,2022.

[12] A. Selim, S. Kamel, A. S. Alghamdi, and F. Jurado, "Optimal Placement of DGs in Distribution System Using an Improved Harris Hawks Optimizer Based on Single- and Multi-Objective Approaches," *IEEE Access*, vol. 8, pp. 52815–52829, 2020.

[13] Y. Gupta, R. Nellikkath, K. Chatterjee, and K. Doolla, "Volt–Var Optimization and Reconfiguration: Reducing Power Demand and Losses in a Droop-Based Microgrid," *IEEE Trans. Ind. Appl.*, vol. 57, no. 3, pp. 2769–2781, 2021.

[14] Nagendra Singh and Yogendra Kumar, "Multi Objective Economic Load Dispatch Problem Solved by New PSO," *Adv. Electri. Engg.*, vol. 2015, pp.2356–6655, 2015. https://doi.org/10.1155/2015/536040

[15] A. A. A. El-Ela, R. A. El-Sehiemy, and A. S. Abbas, "Optimal Placement and Sizing of Distributed Generation and Capacitor Banks in Distribution Systems Using Water Cycle Algorithm," *IEEE Syst. J.*, vol. 12, no. 4, pp. 3629–3636,2018.

[16] S. Singh and P. Nitin, "Optimal Integration of Multi-type DG in RDS Based on Novel Voltage Stability Index with Future Load Growth," *Evol. Syst.*, vol. 12, no. 4, pp. 981–995,2021.

[17] T. Yuvaraj, K. Rukmani Devabalaji, N. Prabaharan, H. Haes Alhelou, A. Manju, P. Pal, and P. Siano, "Optimal Integration of Capacitor and Distributed Generation in Distribution System Considering Load Variation Using Bat Optimization Algorithm," *Energies*, vol. 14, no. 12, p. 3548,2021.

[18] S. S. Parihar and N. Malik, "Network Reconfiguration in the Presence of Optimally Integrated Multiple Distributed Generation Units in a Radial Distribution Network," *Eng. Optim.*, vol. 56, no. 5, pp. 679–699,2024.

[19] O. M. Neda, "Hybrid Design of Optimal Reconfiguration and DG Sizing and Siting Using a Novel Improved Salp Swarm Algorithm," *Electr. Eng.*, pp. 1–16, 2024.

[20] N. Singh, T. Chakrabarti, P. Chakrabarti, M. Margala, A. Gupta, S. B. Krishnan, and B. Unhelkar, "A New PSO Technique Used for the Optimization of Multiobjective Economic Emission Dispatch," *Electronics*, vol. 12, no. 13, p. 2960,2023. https://doi.org/10.3390/electronics12132960

[21] J. Wang, A. Raza, T. Hong, A. C. Sullberg, F. de León, and Q. Huang, "Analysis of Energy Savings of CVR Including Refrigeration Loads in Distribution Systems," *IEEE Transactions on Power Delivery*, vol. 33, no. 1, pp. 158–168, 2018. doi: 10.1109/TPWRD.2017.2710147
[22] N. Singh, A. Nighoskar, and A. Nagar, "New PSO Techniques for the Optimization of Economic Load Dispatch of Thermal Generated Power," *Int. J. Eng. Technol.*, vol.7, no. 2.16, pp. 73–76, 2018. https://doi.org/10.14419/ijet.v7i2.16.11419
[23] N. S. Lakra and B. Bag, "Energy Management in Distribution System Using Volt-VAr Optimization for Different Loading Conditions," *Process Integr. Optim. Sustain.*, vol. 6, no. 2, pp. 295–306, 2022.
[24] S. Chanda, F. Shariatzadeh, A. Srivastava, E. Lee, W. Stone, and J. Ham, "Implementation of Non-intrusive Energy Saving Estimation for Volt/VAr Control of Smart Distribution System," *Electr. Power Syst. Res.*, vol. 120, pp. 39–46, 2015.
[25] V. Veera, V. Satya, N. Murty, and A. Kumar, "Optimal DG Integration and Network Reconfiguration in Microgrid System with Realistic Time Varying Load Model Using Hybrid Optimisation," *IET Smart Grid*, vol. 2, no. 2, pp. 192–202, 2019.
[26] J. M. Home-Ortiz, R. Vargas, L. H. Macedo, and R. Romero, "Joint Reconfiguration of Feeders and Allocation of Capacitor Banks in Radial Distribution Systems Considering Voltage-dependent Models," *Int. J. Electr. Power Energy Syst.*, vol. 107, pp. 298–310, 2019.
[27] V. B. Pamshetti, S. Singh, and S. P. Singh, "Combined Impact of Network Reconfiguration and Volt-VAR Control Devices on Energy Savings in the Presence of Distributed Generation," *IEEE Syst. J.*, vol. 14, no. 1, pp. 995–1006, 2020.
[28] V. V. V. S. N. Murty and A. K. Sharma, "Optimal Coordinate Control of OLTC, DG, D-STATCOM, and Reconfiguration in Distribution System for Voltage Control and Loss Minimization," *Int. Trans. Electr. Energy Syst.*, vol. 29, no. 3, pp. 1–27, 2019.
[29] N. Singh, M. Praveen Kumar, and B. Santosh Kumar, "Effect of Valve Loading on the Thermal Power Economic Load Dispatch Using New Elephant Herding Optimization," *Int. J. Recent Technol. Eng.*, vol. 7, no. 6S4, pp.345–349, 2019.
[30] S. A. Taher and M. H. Karimi, "Optimal Reconfiguration and DG Allocation in Balanced and Unbalanced Distribution Systems," *Ain Shams Eng. J.*, vol. 5, no. 3, pp. 735–749, 2014. https://doi.org/10.1016/j.asej.2014.03.009
[31] N. Singh, T. Chakrabarti, P. Chakrabarti, V. Panchenko, D. Budnikov, I. Yudaev, and V. Bolshev, "Analysis of Heuristic Optimization Technique Solutions for Combined Heat-Power Economic Load Dispatch," *Appl. Sci.*, vol. 13, no. 18, p. 10380, 2023. https://doi.org/10.3390/app131810380

Chapter 4

The role of fiber optics in power system protection

Dipti Shukla

4.1 INTRODUCTION

Protection of electrical systems is the focus of the potent industrial field known as *power system protection*. The goal of this plan is to maintain the stability ofthe electricity system and ensure the network remains stable while it operates. Figure 4.1 shows the schematic diagram of the protection of the power system. Four distinct components make up the protection systems, as shown in Figure 4.1:

- *Voltage and current transformer (CT):* Measuring devices include the potential transformer (PT), sometimes referred to as the voltage transformer, and the CT. A CT reduces the current signals for measurement purposes, whereas a PT converts high-voltage values into lower ones. The transformers are designed to measure electrical systems' precision and safety.
- *Protective relay:* A protective relay is a type of relay used in electrical engineering that is intended to trip a circuit breaker upon detection of a defect.
- *Circuit breaker:* By definition, a circuit breaker is an electrical safety device—a switch that, in the case of an overload, short circuit, or ground fault, immediately cuts off the current. Circuit breakers "trip" or stop allowing current to flow, when a protective relay senses a problem.
- *Batteries:* To supply power in case of a disconnect.

The primary objective of protection in the power system is to isolate the problematic area from the remainder of the framework so that it can continue to function normally. Moreover, it serves to safeguard the electrical system and stop the fault current from flowing. By rapidly cutting off the short circuit, it can aid in stopping the flow from continuing. On the other hand, protection relays are a crucial component of power system protection

 DOI: 10.1201/9781003516156-4

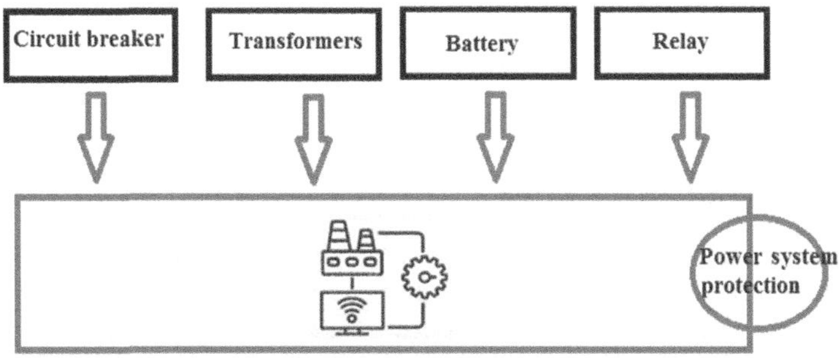

Figure 4.1 Schematic diagram of protection of power system.

that aids in isolating the electrical system's faculties. Still, it is critical that this relay has the following attributes, which are discussed below.

4.2 TYPES OF PROTECTION

4.2.1 Network of high-voltage transmission

Two purposes for protecting the transmission and distribution system are plant safety and public safety (including worker safety). Protection works by, in essence, cutting off power to tools that have an excess or little to earth. Transformers and other components in substations may need extra protection depending on several factors, including gas pressure and temperature.

4.2.2 Sets of generators

Relays that protect a power plant are built to shield alternators from harm or transformers in the case of unusual operating situations, which can arise from internal failures, insulating failures, or regulation problems. Because these kinds of failures are uncommon, the protective relays must be used extremely infrequently. In addition, the potential for expensive equipment repairs or shield to the transformer or alternator may result from a protective relay failing to identify a flaw, leading to replacement owing to the incapacity to generate and trade energy.

4.2.3 Overload and backup (overcurrent) for distance

A CT, which only monitors the current in a circuit and compares it to a preset value, is necessary for overload protection. Overload protection comes in two types: temporal overcurrent (TOC) and instantaneous overcurrent (IOC). For an IOC circuit breaker to activate, the current must be above a predefined threshold. A current vs. time curve serves as the foundation for time overcurrent protection. This curve indicates that the circuit breaker or fuse will trip if the measured current exceeds a certain threshold for the predetermined period.

4.2.4 Ground/earth fault

CTs are also necessary for earth fault prevention, which detects a difference in a three-phase circuit. The three-phase currents are normally balanced and of comparable magnitudes. A low-impedance connection connecting one or both phases to earth will cause a sharp rise in both phase magnitude and current imbalance. A circuit breaker should activate if this imbalance is greater than a set value. Earth fault prevention is limited to detecting earth faults involving two sets of transformers that carry current.

4.2.5 Distance (relay of impedance)

Both voltage and current are detected via distance protection. A circuit failure often causes the voltage level to drop. The circuit breaker will activate if the voltage to current ratio—which is equivalent to impedance—measured at the relay terminals falls within a set range. As their operational properties are dependent on the parameters of the line, this is helpful for relatively long lines—that is, more than 10miles. This implies that the relay's impedance setting is measured and compared to the apparent impedance of the line whenever a fault arises on the line, starting at the relay terminals. The problem is found to be inside the zone of protection if it is discovered that the relay setting is less than the perceived impedance. It is more challenging to coordinate distance protection when the gearbox line is too short (> 10 miles). Current differential protection is the most suitable option for protection in these situations.

4.2.6 Reverse

The goal of protection is to remove the plant just to the extent that it is harmed, leaving the rest intact. It is possible for a protective relay or circuit breaker to fail. When the primary protection of a vital system fails, backup protection frequently activates. Remote backup protection often fixes the problem by removing both the afflicted and unaffected plant components. The affected plant components will be removed via local backup protection to resolve the problem.

4.2.7 Networks with low voltage

To eliminate overload and ground problems, the low-voltage network often utilizes fuses or low-voltage circuit breakers.

4.2.8 Online safety

Every day, there are new cyber-security dangers to the bulk system, a vast network of related electrical systems that includes a control and transmission system—"Cyber-security for the Electric Grid," 2019. These attacks mostly go against the control systems of the grids. Because there is an internet connection to these control systems, they are more susceptible to hacking attempts. These attacks may reduce the authority of utility personnel over the system and may result in equipment damage.

4.3 FIBER OPTICS'S BENEFITS FOR POWER AND ENERGY

Modern communication systems usually employ fiber optic cables, which are sophisticated and versatile network links that carry data across several glass or plastic strands. Fiber optics offers several significant advantages in energy and electricity, but it is also necessary for internet service providers to give more capacity and quicker transmission rates.

To provide dependable electricity to homes and businesses in the safest manner possible, utility providers must overcome several obstacles, which is made possible in several ways by fiber optic connections.

Data is sent via fiber optic cables; electricity is not transferred. Nevertheless, utilities may employ fiber optics to upgrade and modernize energy infrastructure, increasing their safety, efficiency, and modernity. Upgrading to smart grids, which utilize electricity more wisely and depend more on renewable energy to power the system, is one of the most creative approaches.

Fiber optic networks offer a safe and effective means of dividing up data traffic and sending information between the many vital parts of a smart grid, such as sensors, controllers, information technology (IT)systems, measurement instruments, and more. Implementing fiber optics can benefit utilities by improving operations, addressing safety issues, improving customer service, and opening up new revenue streams that are innovative and sustainable.

4.3.1 Sustainability

Sustainability becomes increasingly feasible as more fiber optic and power firms collaborate to convert energy infrastructure into smart grids, which is advantageous to both businesses and consumers. Several factors make fiber optics more environmentally friendly.

4.3.2 Energy efficiency

Fiber optic cables are more effective and have a lower carbon footprint than copper cables because they use less energy to transport data.

4.3.3 Fewer resources

Compared to alternatives, fiber optics use fewer resources. Utility companies require fewer resources to repair or replace fiber optic cables due to their longer durability and resistance to weathering.

4.3.4 Less infrastructure is required to reduce signal loss

Signals carried on copper cables, for example, are prone to deterioration over extended distances, necessitating the use of signal repeaters and other supplementary equipment.

4.3.5 Reduced technological waste

Although technology has completely changed the world, updates and replacements come with a price. The effects of electronic trash, or "e-waste," on the environment and public health are a cause for increasing worry. Fiber optic connections can combine services and cut down on technological waste in situations where providers previously utilized separate lines for various services.

4.4 FIBER OPTIC SENSORS AND APPLICATIONS OF FIBER OPTICS IN POWER SYSTEM PROTECTION

Several fiber optic technologies are characterized by many similar features, including being small, lightweight, protected from electromagnetic interference (EMI), and capable of withstanding high temperatures and pressures [1]. Moreover, fiber optic technology is adaptable and has a wide range of physical quantity measurements. Owing to these benefits, fiber optic sensors are extensively used in the field and are capable of meeting a sizable amount of the present and future needs for power system applications [2].

Figure 4.2 depicts the main sub-systems and mechanism of the power system are represented in this section along with the current state of fiber optic-based sensor technologies utilized in these various applications and the requirements of sensors.

4.4.1 Application of fiber optics sensors in power transformers protection

The focus of this field is fluid-drenched transformers, which are the most often used advancement in transmission and scattering frameworks. Sensors in transformers are mostly used to screen the dielectric quality of the encasing. For the duration of the transformer's lifespan, the security liquid in transformers undergoes changes due to precursors to heat, material,

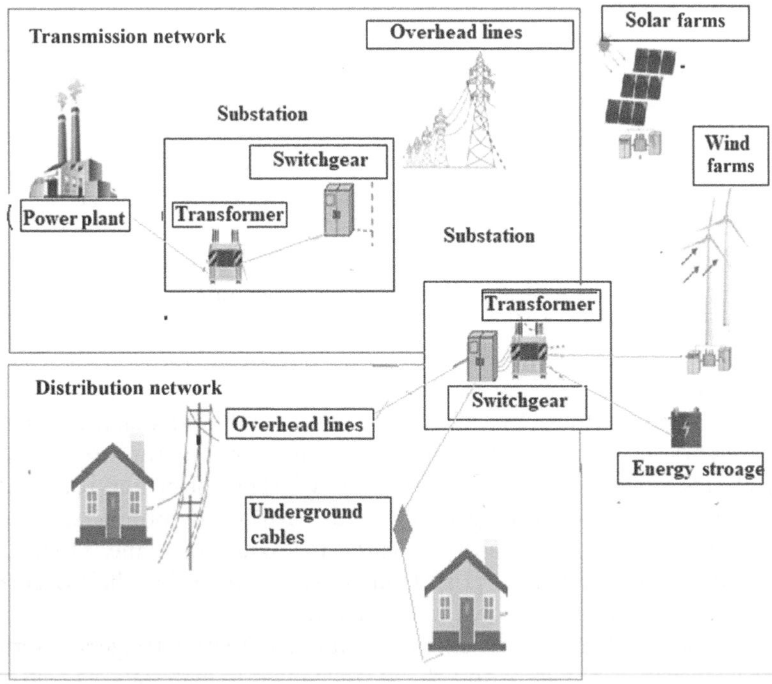

Figure 4.2 Main sub-systems and mechanism of power system.

mechanical, electromagnetic, and electrical loads in the middle of routine and transient stacking settings.

This approach, referred to as separator growing, continuously induces the dielectric quality of the protector to decrease, until total cover breakdown takes place. This process frequently determines a transformer's lifespan. The state of the transformer's stacking and warm stretch are the determining factors in the corrosion of the transformer cover liquid during operation [3]. To anticipate the failure of the control electrical system, a few limits can be observed to identify the degree of transformer corruption: temperature, clamminess, twisting mishappening, and incomplete delivery (PD).

The transformer's most noticeable temperature, referred to as the problem area temperature (**hot spot temperature** (HST)), has a significant influence on cover deterioration. In addition, the HST limits the transformer's stack limit; thus it is important to select both its region and respect. To examine HST in transformers, there are two basic systems to consider: either from the interior using fiber optic sensors or from the exterior using thermocouple devices. Thermocouple devices are susceptible to EMI and are required to estimate the insulating liquid temperature at the bottom or top of the transformer. The usage of fiber optic sensors provides far higher reliability.

As discussed in "Fiber Optic Sensors," there are two fundamental categories of fiber optic sensors that can give multi-point detection: conveyed and quasi-distributed. The conveyed detecting framework, which can measure temperatures at various areas along an optical fiber, utilizing Raman and Rayleigh diffusing, has the disadvantage of poor spatial determination, long securing time, and higher costs [4–7]. These restrictions are overcome utilizing quasi-distributed strategies based on the Fiber Bragg Grating (FBG) sensors, which can be found within the key positions concurring with the real requirements [8–12].

Halfway release (PD) can happen inside separators due to the electrical breakdown of the gas, as previously discussed, in a non-uniform electric field or within a void [13]. They are both a result and a sign of separator framework deterioration in control transformers. Warm stretch, caused by electrical losses (i.e., warming) in the windings or center, breaks down the protection fluid, leading to PD and maturing by-products, including moisture-dissolved gases such as carbon dioxide, methane, ethane, etc., as well as acids and slime [14]. Consequently, observing their occurrence serves as a tool to assess separator conditions.

The sudden discharge of energy during a PD event triggers a series of impacts, such as auxiliary and chemical changes in materials, as well as the generation of electromagnetic signals and acoustic emissions(AEs). These impacts are utilized in identifying PDs, with strategies based on current acceptance [15], ultra-high recurrence [16], and acoustic emission [17] estimations having been created for this reason. Among the diverse types of sensors, the winding is a mishappening. The high current produced by lightning strikes or short circuits can cause mechanical distortions to the windings of transformers that can compromise their proper working.

Fiber optic detecting strategies, including vibration or weight sensors based on nearby or disperse destinations, can distinguish these distortions. For illustration, Reference [18] have created a lean fiber optic sensor utilizing the Fabry–Perot interferometer guideline, which can straightforwardly measure winding vibrations with a transfer speed of 20 to 1000 Hz. FBG sensors are ideal for winding distortion observation because of their capacity to measure different focuses, illustrated by their FBG weight sensor for pivotal constrain estimation of transformer windings [19].

Dampness within the transformer tank can speed up the maturing process of the paper insulation by diminishing its degree of polymerization and atomic weight. Additionally, dampness in combination with high oxygen levels can quicken the maturing rate by 8 to 9 times. These impacts can cause a decrease in mechanical flexibility, dielectric strength, and bubble initiation temperature, as well as an increment in PD.

Subsequently, to move forward with the unwavering quality of control transformers, it is imperative to precisely decide the water content inside the paper separator (and the encompassing fluid). Optical fiber-based sensors, like FBG or evanescent-based wave, are developed for near the beginning location within the changes in dampness concentration. One of the foremost broadly utilized fiberoptic-based sensors for this application is an FBG sensor coated with a moisture-sensitive layer. For example, a polyimide layer on an FBG can measure water content in transformer cover, achieving a sensitivity of less than 1% relative humidity and an affectability of 0.9 pm/1% RH. Alternatively, the utilization of the poly-methyl methacrylate (PMMA) as moisture-sensitive material improves the sensor affectability (RH=29 pm/1%). Moreover, a fiber optic detecting gadget for the synchronous observing of temperature, vibration, and Fabry–Perot depression-based dampness was made using two FBGs.

However, evanescent-wave sensors, particularly FBG-based ones, have advanced in terms of temporal response and are less sensitive to changes in temperature. The inability to read the humidity data contained in the signal's concentrated area is an evanescent-wave sensor's drawback. Because of variations in the optical fiber location or the noise from the light source, the intensified light is more susceptible to irregular changes in wavelength (as a twist in the fiber might induce a concentrated constriction).

For example, cutting an optical fiber at an angle to create a location near the fiber center where there is a strong interaction with the light is one of the many methods for obtaining an evanescent-wave sensor. Another approach to forming an interaction with the light engendering within the fiber comprises unequivocally twisting a plastic optical fiber causing light from the center to enter the cladding fiber (an external portion of the optical fiber) that is in contact with the outside environment with which it can connect in turn [20]. Fiber optic innovation is increasingly being utilized for acoustic signal estimations.

4.4.2 Application of fiber optics sensors in overhead line protection

Using sensors to measure variations in temperature, strain, and vibration along the control lines, sensors can identify trouble on the line that may indicate possible problems like listing lines or damaged components. This enables administrators to mandate preventative actions to avert blackouts and other problems. Additionally, sensors may help advance security by warning utilities about possible security risks like downed or broken control lines, predicting mishaps, and lowering the possibility of harm to experts and the community.

Furthermore, proximity to sensors is required to advance the functionality of the control systems by giving real-time data on the state of the control lines. This may be used to advance resource employment through plan operation and assist managers cut down on the number of manual evaluations needed to optimize their maintenance plans, which saves money and time. For instance, optical fiber innovation also contributes to a successful and lucrative setup (mostly due to their ability to observe distinct parameters on the overhead control lines, reach great distances, survive harsh environmental conditions, and provide more precise and timely information).

4.4.2.1 Temperature

A conductor's temperature is the main factor affecting its capacity to carry current. Temperature calculations throughout the line are therefore crucial in measuring the energy hot rating (DTR) of visual projection wires. DTR frameworks employ temperature sensors to continually measure the temperature of the gearbox line and supply real-time data for calculations that establish the current-carrying capability. Conveyance detection based on Brillouin and Rayleigh diffusing within the fiber optics is the most commonly utilized innovation for temperature screening along control lines (because of their accurate tall area, rapid reaction time, and dependable operation execution). DTR frameworks provide the organization administrator flexibility by giving them access to real-time data about the system's capacity to bear loads. Future systems are anticipated to rely more and more on these frameworks. More details about them are provided in the "Asset Utilization" section.

Lightning is a crucial component that must be considered to guarantee the security and dependability of the control framework. Finding the incidence, recurrence, and harshness of lightning strikes is crucial for preservation operations, even though unusual parts like lightning arrestors, which essentially reroute lightning and restrict engendering into the plan, help to shield the control lines and limit harm from strikes.

Fiber optic sensors are a useful instrument for locating lightning strikes in overhead control lines because they can function in harsh environments and are unaffected by interference from other electrical signals. They offer extremely accurate and definitive information on the strike's location and strength. These sensors, which are often dispersed and placed specifically on the control lines, are sensitive to variations in electromagnetic fields caused by lightning. One may ascertain the modifications to the optical fiber's refractive index brought about by the electromagnetic fields generated by a lightning strike by measuring differences in the light that passes through the fiber.

4.4.2.2 Icing

The icing process occurs when there is snow, low relative humidity, and freezing temperatures. Control line disappointments such as insulator surface flashover, line disengagement, and tower disappointment can be caused by an accumulation of ice on the conductor surface. The ice mass may be calculated using a few different techniques:(1) climatological model forecasting, (2) visual inspection with various cameras positioned at key points, (3) capacitive sensors installed on the conductors specifically, and (4) stack cells, which use the tension between the separators property of the conductor and the electric tower arm to assess the quantity of ice present. Many fiber optic sensors used in strain measurements belong to the latter kind (load cell) due to the practicality and effectiveness of FBGs.

4.4.2.3 *Line dashing*

An oscillating movement happens when visual projection lines are subjected to high winds or other external conditions, like ice or blizzard accumulation. A major push might cause damage or perhaps the collapse of the power line towers if the motion becomes too intense. Higher-voltage parallel transmission lines are extra susceptible to this as higher towers are where they are situated. Owners of frames must thus take several safety measures to lower the danger, such as vibration-resistant designs, companion wires, and dampers. However, using sensors to filter the phenomenon is equally crucial.

4.4.2.4 *Separator checking*

Separators need to be inspected in electrical energy transmission networks to guarantee maximum efficiency and safety. Monitoring insulin leakage current is also necessary. An attractive toroid surrounding the covers, an acceptance coil, or a resistive impact sensor are used in electrical framework devices. Numerous fiber optic sensors gradually transform the calculated current back into a light intensity reading. In particular, they modify a light emitting diode (LED) to produce light that is fixed to the tower and covers.

4.5 APPLICATION OF FIBER OPTIC SENSORS IN UNDERGROUND CABLE PROTECTION

Subterranean electrical cable frameworks primarily employ sensors to continuously ensure the greatest levels of efficiency and security by monitoring temperature and separation conditions. This could be required because, although the framework is shielded from most externally caused damage and defects, in comparison to overhead line frameworks, repair times might still be up to 10 times longer. Sensors are installed at standard intervals throughout the total wire length to enable permanent monitoring and early problem discovery. The literature presents a wide range of fiber optic sensors for PD discovery. In particular, some studies detect the sound waves resulting from PD occurrences using dispersed plans. To distinguish between the vibration resulting from PD occurrences, cable-PD testing approaches based on Rayleigh diffusing were employed. Additionally, a significant number of neighborhood sensors that make use of interferon technology have been shown, including designs that are based on interferometers.

4.6 APPLICATION OF FIBER OPTIC SENSORS IN WIND POWER PLANTS

Sensors are employed in wind turbines to filter a variety of data, including electrical quantities and mechanical/fatigue/stress variables that are related to plant security and generation optimization. Because of the complex construction, a few special sensors are employed; however, fiber optic-based sensors are frequently chosen.

Reinforced concrete structures are impacted by several forms of corruption, including under-reinforcement through design, shrinkage, warm angles through casting, and freeze–thawing through installation. By utilizing the right sensors, which provide the capability of viewing the underground breaks in reality without halting the turbines, the limitations of visual examinations(schedule checks) may be overcome. Perry and McAlorum, for example, showed that it is possible to use FBG sensors for break observations during wind turbine assembly.

The wind turbine edges are another component that has to be protected because of the strong forces that are operating on them. These forces raise the possibility of fatigue disappointment and may even reduce turbine power. In actuality, several edge-checking analyses using fiber optic sensors, more specifically FBGs, have also been carried out.

Basic testing of underwater control cables is also necessary while operating at sea. Apart from the elements covered in "Underground Cables," the surroundings of the facility also require careful inspection as the underwater cables may be subjected to high sales. In addition, they have to function under harsh conditions involving water pressure, salt, sea currents, and other factors that might destroy copper conductors. Authors of [6] performed a numerical study of sales in a control cable to optimize the sensor areas, and they planned a technique for the control cable observation based on FBG sensors.

In terms of the electrical parts, it has been found in later applications that fiber optic sensors can, for instance, employ FBG to observe the temperature within the semiconductor electronic devices applied in control electronic modules. These have led to the creation of a dispersed warm checking system that employs FBG to regulate devices in wind turbines as they have the potential to significantly contribute to device problems. Furthermore, FBG sensors may be connected to track the temperature and mechanical properties (strain, vibration, torque, and rotation speed) of electrical devices that rotate. These variables are all essential for condition monitoring.

4.7 USE OF FIBER OPTICS-BASED SENSORS IN SOLAR FORM PROTECTION

The quantity of sun-based energy retained and the temperature both affect the manufacturing of photovoltaic cells. These need to be adjusted appropriately to guarantee that the framework functions as best it can. Investigating the temperature outline and ensuring adequate cooling within the solar modules is crucial for advancing the efficacy of the control era. For conventional sensors, FBGs' distinct characteristics—such as their natural multiplexing, EMI resistance, erosion resistance, minimum estimate, and so forth—have demonstrated that they are a very useful invention in this business. For this reason, the previous research introduced two FBG-based temperature detection techniques.

Verifying the temperature of the solar board's surface is also crucial for preventing the hotspot problem. It comprises a steady increase in temperature inside the faulty areas of the board, influencing the lifespan and performance of the power generation process. The authors described a detailed procedure for locating the hotspot with FBG sensor clusters..

4.8 APPLICATION OF FIBER OPTIC SENSORS IN ENERGY STORAGE

Battery monitoring frameworks have two basic goals: to get data on the state of health (SOH) and state of charge (SOC) of batteries. For battery observation, many fiber optic-based detection methods have recently been developed. In particular, utilizing temperature estimations, a range of fiber optic sensors, including FBGs and Rayleigh diffusing sensors, have been suggested to ascertain safe and dependable battery operation. Improper cooling can indeed result in disappointments, fires, or even explosions from batteries. Temperature estimations can be performed within the battery to provide more accurate data on its state or very simply on its outside surface with a large gain in framework complexity. Thus, at the moment, internal estimating frameworks are restricted to research-scale activities.

The cathode push, which affects battery capacity and management, is another essential indicator for battery observation. Strain sensors, such as FBGs placed on the terminals, may easily detect this parameter. This allows for a strategic separation from more sophisticated advancements, such as advanced picture relationships, or X-ray diffraction.

4.9 APPLICATION OF FIBER OPTIC SENSORS IN SWITCHGEAR AND PROTECTION

Sensors in switchgear are mostly used in the PD movement check process to provide useful detail for the separator position of the control devices and to predict malfunctions. The most often used devices for this assignment at the moment are electrical sensors, which can be broadly divided into two groups: (1) sensors that follow the standard IEC 60270, which carries out PD estimations in a recurrence span less than 13 × 10^6Hz; and (2) sensors that follow the standard IEC 62478, which carries out PD estimations at frequencies greater than 3 ×10^6Hz.

In any case, the preferences of the optical fiber sensors discussed in previous sections make this innovation increasingly alluring. The fiber optic sensors for the PD location are based on interferometry strategies, such as Sagnac interferon and Michelson interferon. Another approach described utilizes the optical triangulation strategy to gauge the contact quality utilizing the pictures given by two micro cameras.

The control framework provides current and/or voltage predictions that are necessary for some switchgear devices, including transfers, to function. Current or voltage transformers are typically used to get these estimates; however, because of the transformer center's immersion and magnetism, they are susceptible to error. Various other fiber optic sensor-based methods, including FBG and Faraday pivoting sensors, have been suggested.

4.10 CONCLUSION

The addition of multi-objective carbon advances and more powerful control framework operation are key mechanisms within the move to a maintainable prospect. Following these improvements is an ever-expanding deployment of sensors, which assist control framework administrators in arranging, processing, and maintenance tasks. Due to their ability to measure a few characteristic physical parameters, fiber optic sensors are recognized as acritical empowering novelty and offer numerous opportunities to enhance situational mindfulness in organized frameworks. These applications have been investigated in this chapter, which has displayed a comprehensive summary of the role of fiber optic sensors in control frameworks.

REFERENCES

1. Singh, N. Application of fiber optics for the protection and control of power systems, In: *Optical to Terahertz Engineering. Progress in Optical Science and Photonics*, vol.23, pp.101–120, 2023. Springer, Singapore. https://doi.org/10.1007/978-981-99-0228-6_7
2. Singh, N., Chakrabarti, T., Chakrabarti, P., Panchenko, V., Budnikov, D., Yudaev, I., Bolshev, V. Analysis of heuristic optimization technique solutions for combined heat-power economic load dispatch. *Applied Sciences*,13(18), 10380, 2023. https://doi.org/10.3390/app131810380
3. Wieduwilt, T., Förster, R, Nissen, M., Kobelke, J., Schmidt, M.A. Characterization of diffusing sub-10 nm nano-objects using single anti-resonant element optical fibers. *Nature Communications*,14, 3247,2023.

4. Pendao, C., Silva, I. Optical fiber sensors and sensing networks: Overview of the main principles and applications. *Sensors*, 22(19), 2022.
5. Swain, A., Abdellatif, E., Mousa, A., Pong, P.W.T. Sensor technologies for transmission and distribution systems: A review of the latest developments. *Energies*,15(19), 2022.
6. Fook, V.F.S., Jayachandran, M., Jiliang, E.P.,Yongwei, Z., Jianzhong, E.H. FiberBragg grating-based monitoring and alert system for care of residents in nursing homes. In: *2018 IEEE 4th World Forum on Internet of Things (WF-IoT)*, pp. 195–200, 2018. IEEE.
7. Singh, N., Tirole, R., Almas, S.K., Sood, D. Renewable power generation using asynchronous generator: A review, In: *Recent Advances in Power Systems*. Lecture Notes in Electrical Engineering, vol.699, pp. 205–215, 2021. https://doi.org/10.1007/978-981-15-7994-3_18
8. Aldaba, A.L., González-Vila, Á., Debliquy, M., Lopez-Amo, M., Caucheteur, C., Lahem, D. Polyaniline-coated tilted fiber Bragg gratings for pH sensing. *Sensors and Actuators B: Chemical*, 254, 1087–1093, 2018.
9. Singh, N., Priyadarshi, N., Kumar, N. Meta heuristic techniques used to improve the performance of static converters, In: *International Conference on Innovative Development in Virtual and Rapid Prototyping (IDVRP-2022)*. AIP Conference Proceedings 2771, 020027, 2023. https://doi.org/10.1063/5.0152913
10. Wild, G., Hinckley, S. Acousto-ultrasonic optical fiber sensors: Overview and state-of-the-art. *IEEE Sensors Journal*, 8(7), 1184–1193, 2008.
11. Zhou, H.-Y., Ma, G.-M., Wang, Y., Qin, W.-Q., Jiang, J., Yan, C., Li, C.-R., Optical sensing in condition monitoring of gas insulated apparatus: A review. *High Voltage*, 4(4), 259–270, 2019.
12. Islam, M., Ali, M.M., Lai, M.-H., Lim, K.-S., Ahmad, H. Chronology of Fabry–Perot interferometer fiber-optic sensors and their applications: A review. *Sensors*,14(4), 7451–7488, 2014.
13. Veena, R.S., Khambra, G., Sayal, R., Singh, N., et al. An energy efficient IPv6 packet delivery outline for industrial IoT over G.9959 protocol based wireless sensor network (WSN). *SN ComputerScience*, 5, 813, 2024. https://doi.org/10.1007/s42979-024-03195-4
14. Palma, P.D., Natale, D., Campopiano, S., Iadicicco, A.A simple Fabry–Perot pressure sensor fabricated on fiber optic tip. In: *Sixth European Workshop on Optical Fibre Sensors*, 9916, 138–141, 2016.
15. Ma, J., Xuan, H., Ho, H.L., Jin, W., Yang, Y., Fan, S. Fiber-optic Fabry–Perot acoustic sensor with multilayer graphene diaphragm. *IEEE Photonics Technology Letters*, 25(10), 932–935, 2013.
16. Chen, D., Qian, J., Liu, J., Chen, B., An, G., Hong, Y., Jia, P., Xiong, J. An in-line fiber optic Fabry–Perot sensor for high-temperature vibration measurement. *Micromachines*, 11(3), 252, 2020.
17. Almas, S.K., Singh, N. Power quality improvement for grid connected wind energy. Generation system by implementation of DFIG scheme. *Turkish Online Journal of Qualitative Inquiry (TOJQI)*, 12(9), 6928–6944, 2021.
18. Wei, P., Shan, X., Sun, X. Frequency response of distributed fiber-optic vibration sensor based on nonbalanced Mach–Zehnder interferometer. *Optical Fiber Technology*, 19(1), 47–51, 2013.

19. Hong, X., Wu, J., Zuo, C., Liu, F., Guo, H., Xu, K. Dual Michelson interferometers for distributed vibration detection. *Applied Optics*, 50(22), 4333–4338, 2011.
20. Wu, D., Zhu, T., Liu, M. A high temperature sensor based on a peanut-shape structure Michelson interferometer. *Optics Communications*,285(24), 5085–5088, 2012.

Chapter 5

Optimizing thermal power plant performance

A comprehensive energy audit

Deepak Rathod, Ankit Tiwari, Lata Gidwani, and Alka Karketta

NOMENCLATURE

Abbreviation	Description
AHP	air preheater
CEP	condensate extraction pump
DA	deaerator
DM	demineralization
FD	forced draft
HP	high pressure
ID	induced draft
IGV	inlet guide vane
LDB	lighting distribution board
LP	low pressure
MU	mega units
NA	not available
SAC	self-adjusting clutch
SPC	specific power consumption
VFD	variable frequency drive

5.1 INTRODUCTION

An energy audit of a thermal power plant examines the boiler, electrical systems, pumping equipment, air compressor, cooling towers, auxiliary power use, and so on [1,2]. This study evaluates a 100-MW thermal power facility's energy efficiency, proposing potential energy-saving solutions for factories, aiming to reduce monthly electrical energy consumption and manufacturing expenses through comprehensive energy surveys. This energy analysis includes essential information such as the factory's energy consumption patterns, the identification of energy-intensive machinery, and potential energy-saving strategies [3].

DOI: 10.1201/9781003516156-5

5.2 SCOPE OF AUDIT

An energy audit examines various aspects of a coal handling plant, including auxiliary power consumption, raw water intake, demineralization (DM) plant, ash handling plant, feed water systems, steam systems, secondary fuel oil consumption, vacuum system, plant and control air system, turbine heat rate, ball mills, air fans, cooling towers, condensation extraction system, plant lighting, transformers, and generator efficiency evaluation [4].

5.3 CONDENSATE EXTRACTION PUMPS

The generating unit has three 132 kW condensate extraction pumps (CEPs), with two in full generation. Electrical parameters, flow, and pressure were measured for combined efficiency [5]. Electrical parameters, flow, and pressure taken during the site measurement are tabulated in Tables 5.1 and 5.2.

Table 5.1 CEP analysis for transformer 1

S. no.	*Parameters*	*CEP # B*	*CEP # C*
A.	**Design parameters**		
1	Name of plant	100MW	
2	Make and type	BHEL, 150C JNV/4	
3	Rated capacity	142 m^3/h	
4	Speed (rpm)	1485	
5	Head (m)	140	
6	Pump efficiency	NA	
7	Motor rating (kW)	132	
B.	**Measured parameters**		
8	Voltage (volts)	409	408
9	Load current (ampere)	156	168
10	Power factor	0.85	0.85
11	Input power (kW)	93.3	101
12	Frequency (Hz)	49.2	49.2
13	Flow rate (m^3/h)	240	240
14	Velocity (m/sec)	1.934	1.934
15	Suction/discharge pressure (kg/cm^2)	–0.4/12.5	–0.4/13.5
C.	**Performance calculations**		
16	Load (%)	66	72
17	Differential pressure (m)	134	134
18	Calculated pump output (kW)	87.6	87.6
19	Combined input power to the motors	194.3	
20	Motor shaft power (kW)	88	95
21	Combined overall efficiency (%)	45	
22	Computed motor efficiency (%)	94	94
23	Computed efficiency of the pump (%)	48	48
24	Throttling (%)	100	100

Table 5.2 CEP analysis for transformer 2

S. No.	*Parameters*	*CEP #A*	*CEP # B*
A.	**Design parameters**		
1	Name of plant	100MW	
2	Make and type	BHEL, 150C JNV/4	
3	Rated capacity (m^3/h)	142	
4	Speed (rpm)	1485	
5	Head (m)	140	
6	Pump efficiency	NA	
7	Motor rating (kW)	132	
B.	**Measured parameters**		
8	Voltage (volts)	432	412
9	Load current (ampere)	164	150
10	Power factor	0.83	0.85
11	Input power (kW)	102	90.2
12	Frequency (Hz)	49.6	49.6
13	Flow Rate (m^3/h)	250	250
14	Velocity (m/sec)	2.015	2.015
15	Suction/discharge pressure (kg/cm^2)	–0.4/12.6	–0.4/12.5
C	**Performance calculations**		
16	Load (%)	73	64
17	Differential pressure (m)	129.5	129.5
18	Calculated pump output (kW)	88.2	88.2
19	Combined input power to the motors	192.2	
20	Motor shaft power (kW)	96	85
21	Combined overall efficiency (%)	46	
22	Computed motor efficiency (%)	94	94
23	Computed efficiency of the pump (%)	49	49
24	Throttling (%)	100	100

The combined efficiency obtained for the CEPs is 46%. The lower efficiency is due to throttling operations at the flow control valve to maintain the deaerator (DA) level [6]. An energy audit revealed fluctuating plant load due to lignite quality and induced draft (ID) fan issues, causing throttling and energy loss. Installing variable frequency drives (VFDs) could save 20%.

Present average power consumption by 2 CEPs = 192.2 kW
Power saving potential @ 20% = 38.4 kW
Units saved per year = 304,445 kWh
Considering 330 days/year and 24 hours of operation
Amount saved per year @ Rs. 2.0/kWh = Rs. 6.08 lakhs
Approximate investment required for 3 VFDs = Rs. 15.0 lakhs
Payback period = 30 months

5.4 FORCED INDUCED DRAUGHT

5.4.1 Forced draft

The generating unit consists of two forced draft (FD)and two ID fans. The nameplate specifications and electrical parameters measured during the energy audit are tabulated in Table 5.3. Both FD fans are kept in operation to meet the combustion air requirements of the boiler. Airflow measurement of FD fans could not be measured as the same was not approachable. Hence, the airflow, suction, and discharge pressure readings were taken from the control room [7]. Based on measurements and additional data

Table 5.3 Forced draft fan analysis (FD fan)

S.No.	*Parameters*	*FD fan #A*	*FD fan # B*
A.	**Design parameters**		
1	Name of plant	100MW	
2	Make and type	BHEL, NDZV 18 TIEFSTACK	
3	Rated capacity (m^3/sec)	55.6	
4	Fan efficiency (%)	76.40%	
5	Head (m)	373 mm WC at 42°C	
6	Pump efficiency (%)	76.40	
7	Motor rating (kW)	320	
8	Speed (rpm)	980	
9	Medium handled	Air	
B.	**Measured parameters**		
10	Voltage (kV)	6.79	6.81
11	Load current (ampere)	27.8	28.4
12	Power factor	0.71	0.71
13	Input power (kW)	232	238
14	Frequency (Hz)	49.4 Hz	49.4 Hz
15	Flow rate (m^3/h)	53.3 m^3/sec	53.3 m^3/sec
16	Air temperature (°C)	32.5	32.5
17	Suction/discharge pressure (kg/cm^2)	Ambient/229.28*	Ambient/236.17*
C.	**Performance calculations**		
18	Load (%)	67	69
19	Differential pressure (m)	233 mm WC	
20	Density of air is 32.5°C	1.155 kg/m^3	
21	Calculated air power delivered by the fans	122kW	
22	Total input power to the motors(kW)	470	
23	Motor shaft power (kW)	216	221
24	Combined overall efficiency (%)	26	
25	Computed motor efficiency (%)	93	
26	Computed efficiency of the pump (%)	28	
27	Throttling (%)	45	30

* shows the approximate values.

collected during the energy audit, the efficiency of FD fans is calculated, and the detailed calculation is given below.

FD fans A and B's efficiency is 28% below design, due to reduced flow and pressure requirements. Inlet guide vane(IGV) control is effective for modest flow reductions, but energy efficiency drops below 80% flow due to variable plant flow requirements [8]. Hence, it is suggested to install a VFD for both the FD fans, keeping the IGV and damper positions fully open, to fulfill the plant's variable combustion air requirement [9]. The supplier reported a 95% efficiency increase in some plants after installing VFDs, with an additional 70%–80% increase after considering 10% system losses.

Power consumed by FD fans A and B = 470 kW
Power saving potential @ 37% (65 - 28=37) = 174 kW
Units saved per year = 13, 78,080 kWh
Considering 24 hours a day and 330 days/year of operation:
Amount saved per year @ Rs. 2.0/kWh = Rs. 27.5 lakhs
Approximate investment required (for two fans) = Rs. 100 lakhs
Payback period = 43.6 months

5.4.2 Induced draft

The nameplate specifications and electrical parameters measured during the energy audit are tabulated in Table 5.4. Both ID fans are operated to extract the flue gas from the furnace and deliver it to the chimney. Based on measurements and additional data collected during the energy audit, the efficiency of ID fans is calculated, and the detailed calculation is given below.

ID fans A and B have 40% and 60% efficiency, respectively, with lower efficiency due to shaft bend and worn impeller. Proper maintenance and lubrication increase efficiency by 15%–20% [10].

Input power for ID fan A = 334 kW
Power saving potential @ 15% = 50.1 kW
Units saved per year = 3, 96,792 kWh
Considering 24 hours a day and 330 days/year of operation:
Amount saved per year @ Rs. 2.0/kWh= Rs. 7.93 lakhs
Approximate investment required= Rs. 20.0 lakhs
Payback period= 30.3 months

5.5 REDUCTION OF PRESSURE SETTING FOR SERVICE AIR SYSTEM

The service and instrument air compressors discharge compressed air to the common header, but this practice is inefficient due to heat exchanger

Table 5.4 ID fan analysis (FD fan)

S.no.	*Parameters*	*ID fan #A*	*ID fan # B*
A.	**Design parameters**		
1	Name of plant	100MW	
2	Make and type	BHEL, NDZV 25 SIDOR	
3	Rated capacity (m^3/sec)	100	
4	Fan efficiency (%)	78%	
5	Head (m)	324 mm WC at 172°C	
6	Pump efficiency (%)	83	
7	Motor rating (kW)	500	
8	Speed (rpm)	695	
9	Medium handled	Flue gas	
B.	**Measured parameters**		
10	Voltage (kV)	6.81	6.84
11	Load current (ampere)	42.3	53.0
12	Power factor	0.67	0.71
13	Input power (kW)	334	446
14	Frequency (Hz)	49.8	49.7
15	Flow rate (m^3/h)	71	145
16	Gas temperature (°C)	177	176
17	Suction/discharge pressure (kg/cm^2)	−175/4.46	−169/9.4
C.	**Performance calculations**		
18	Load (%)	63	84
19	Differential pressure (m)	180 mm WC	178 mm WC
20	Gas density (kg/m^3)	0.78	
21	Flue gas velocity (m/sec)	9.53	
22	Duct area (m^2)	7.45	
23	Calculated air power delivered by the fans(kW)	125	253
24	Motor shaft power (kW)	355	474
25	Overall efficiency (%)	37.4	56.7
26	Computed motor efficiency (%)	94	
27	Computed efficiency of the pump (%)	40	60
28	Throttling (%)	50	64

inefficiency and frequent scaling. Redesigning the cooling system is recommended [6].

Average power consumption of one air compressor = 104 kW
Power saving potential at 3.0 kg/cm^2 pressure @ 27% = 28.0 kW
Power saving potential for one compressor switching off = 104 kW
Unit saved per year = 10, 45,440 kWh
Considering 330 days per annum and 24 hours/day operation:
Amount saved per year @ Rs 2.00/kWh = Rs. 20.9 lakhs
Investment required = Rs. 1.00 lakhs
Payback period = 0.5 month.

5.6 LIGHTING EFFICIENCY STUDIES

Inside and outside the generating unit, different types of lamps light up different places. The lighting load of the producing unit was measured at night, and the results are shown in Tables 5.1 and 5.2.

5.6.1 Lighting load analysis (night-time)

Reducing lighting supply voltage to 230 volts

The optimal lighting voltage is 230 volts, as higher voltages increase lamp power consumption, reduce burning life, and increase failure rates. To save power, set the tap position of transformers [11].

Night-time

Total lighting load during night-time = 105.64 kW
Supply voltage measured = 241–256 volts
Power saving potential by maintaining 230 volts in the lighting feeder.
(Calculated lighting distribution board (LDB)-wise) = 13.2 kW
Units saved = 57,816 kWh
Considering 365 days and 12 hours of operation per day:

Daytime

Total lighting load during daytime = 30.2 kW
Supply voltage measured = 239–256 volts
Power saving potential by maintaining 230 volts in the lighting feeder.
(Calculated LDB-wise) = 4.36 kW
Units saved = 19,097 kWh
Considering 365 days and 12 hours of operation per day:

Table 5.5 Lighting load analysis (night-time)

S.no.	*Description*	*Location*	*Rating (kVA)*	*Volt*	*Amp*	*P.F.*	*kW*	*kVA*	*Load (%)*	*Remark*
1.	LDB # 1	6.6 kV S/S	100	249	63	0.9	15	16	16	Higher voltage
2	LDB # 2		100	255	49	0.9	12	13	13	
3	LDB # 3		200	242	0	0	0	0	0	
4	LDB # 4		50	244	25	0.9	5.5	6.1	12	
5	LDB # 6		50	242	32	0.5	4.1	7.6	15	
6	LDB # 13		100	242	53	0.9	11	13	13	
7	ELDB # 12		100	243	30	0.7	5.5	7.4	7	
8	LDB # 14		100	242	9.9	0.5	1.3	2.4	2	
9	LDB # 15		100	244	5.5	0.7	0.9	1.3	1	
10	ELDB # 1		100	249	28	1	6.6	6.9	7	
11	LDB # 16		50	256	8.1	0	0.9	2.1	4	

Table 5.6 Lighting load analysis (daytime)

S.no.	Description	Location	Rating (kVA)	Volt	Amp	P.F.	kW	kVA	Load (%)	Remarks
1	LDB # 1	Switch Yard	100	251	57	0.9	13	14	14	Higher voltage
2	LDB # 2		100	256	15	1	3.8	3.9	4	
3	LDB # 3		200	241	5	0.8	0.9	1.2	1	
4	LDB # 4	Turbine Unit	50	244	21	1	5	5.2	10	
5	LDB # 6	Boiler-2 Lighting	50	244	0	0	0	0	0	
6	LDB # 13	B.C.W. Pump House	100	243	2	0.6	0.3	0.5	0	
7	ELDB # 12	D.M. Plant Lighting	100	243	0	0	0	0	0	
8	LDB # 14	C.W. Pump House Lighting	100	239	0	0	0	0	0	
9	LDB # 15	Boiler Lighting	100	239	0	0	0	0	0	
10	ELDB # 1	Service Building Lighting	100	249	11	0.9	2.6	2.8	3	
11	LDB # 16	Switch Yard	50	256	0	0	0	0	0	

Total units saved per annum = 76,913 kWh
Amount saved per year @ Rs. 2.0/kWh = Rs. 1.54 lakh
Approximate investment required = Nil (Tables 5.5 and 5.6).

5.7 CHILLER COMPRESSOR

The AC plant unit's three chiller compressors have a 45% and 146% higher specific power consumption than the design value, requiring maintenance and de-statement [12]. Electrical parameters taken during the site measurement are tabulated in Table 5.7.

Considering the improvement in specific power consumption (SPC) up to 0.90 kW/TR

Power saving potential for Chiller B = 9.5 kW
Power saving potential for Chiller C = 23.8 kW
Total power saving potential = 33.3 kW
Units saved per annum = 2, 91,708 kWh

Considering the 24-hour/day and 365-day/year operation of one chiller

Amount saved per year@ Rs. 2.0/kWh = Rs. 5.83 lakhs
Approximate investment required = Rs. 1.5 lakhs
Payback period = 3.0 months

Table 5.7 Performance of chiller compressor

S. no.	*Parameters*	*Chiller compressor # B*	*Chiller compressor # C*
A.	**Design parameters**		
1	Location	AC plant	
2	Make and type	Voltas, 5H60	
3	Rated capacity (TR)	50	
4	Motor rating (kW)	45	
5	Design specific power consumption (kW/TR)	0.8	
B.	**Measured parameters**		
6	Voltage (volts)	423	423
7	Current (ampere)	71.9	72.9
8	Power factor	0.88	0.89
9	Input power (kW)	46.4	47.5
10	Frequency (Hz)	49.0	49.4
11	Chilled water inlet/outlet temperature (°C)	4.8/0.5	4.8/2.2
12	Chilled water flow rate (m^3/h)	25.8	25.8
13	Chilled water inlet/outlet pressure (kg/cm^2)	1.5/1.2	1.5/N. P
14	Cooling water inlet/outlet Temperature (°C)	30.3/32.5	30.3/32.3
15	Cooling water supply/return pressure (kg/cm^2)	2.5/–0.2 (vacuum)	2.5/–0.2 (vacuum)
16	Chilled water flow rate (m^3/h)	110	120
C.	**Performance calculations**		
17	Load (%)	95	97
18	Chilled water temperature difference (°C)	4.3	4.8
19	Specific heat of water	1.0 kcal/kg°C	1.0 kcal/kg°C
20	Heat load (kcal/h)	110,940 kcal/h	67,080 kcal/h
21	TR output	36.7TR	22.2
22	Computed specific power consumption (kW/TR)	1.16 kW/TR	1.97 kW/TR

5.8 ASH HANDLING PLANT

5.8.1 High-pressure water pumps

The ash plant unit has two 190 kW high-pressure (HP) water pumps for ash handling, with a single pump maintaining the bottom ash removal and air washer systems [9,13,14]. Electrical parameters taken during the site measurement are tabulated in Table 5.8.

HP water pumps 1 and 2 have 66% and 63% efficiency, respectively, but are required for air washer systems during fly ash operations, necessitating a separate pump for energy conservation.

Table 5.8 High-pressure water pump (AHP)

S.no.	*Parameters*	*HP water pump # 1 (fly ash operation)*	*HP water pump # 2 (fly ash operation)*
A.	**Design parameters**		
1	Name of plant	Ash handling plant	
2	Make and type	Beacon Weir, SDC 200/200	
3	Rated capacity	600 m^3/h	
4	Speed (rpm)	1480	
5	Head (m)	80	
6	Pump efficiency	NA	
7	Motor rating (kW)	190	
B.	**Measured parameters**		
8	Voltage (kV)	6.69	6.67
9	Load current (ampere)	16.4	16.2
10	Power factor	0.89	0.9
11	Input power (kW)	169	168
12	Frequency (Hz)	49.2	49.1
13	Flow (m^3/h)	466	450
15	Suction/discharge pressure (kg/cm^2)	0.1/8.4	0.1/8.2
C.	**Performance calculations**		
16	Load (%)	84	84
17	Total head developed (m)	83	81
18	Computed motor efficiency (%)	94	94
19	Calculated pump output (kW)	105	99.3
20	Motor shaft power (kW)	158.6	158
21	Overall pump efficiency (%)	62	59
22	Computed motor efficiency (%)	94	94
23	Computed efficiency of the pump (%)	66	63
24	Design efficiency at duty point of similar rating pump (%)	70	70
25	Throttling (% open)	100	100

Specifications for the new pump for the air washer system:

Flow = 125 m^3/h; head = 60 m; pump efficiency = 70%
Rating of new pump required = 37.0 kW
The estimated input power consumption of the new pump is 31.7 kW.
Present average power consumption by one HP water pump=168kW
Power saving potential = 136.3 kW; units saved = 9, 89,538 kWh
Considering 330 days and 22 hours/day of stopping the existing HP pump during fly ash operation:
Amount saved per year @ Rs. 2.0/kWh = Rs. 19.8 lakhs
Approximate investment required for a 37-kW new pump (2 no.) = Rs. 4.0 lakhs
Payback period = 2.4 months

5.8.2 Low-pressure water pumps

The ash handling plant uses two 35 kW low-pressure (LP) water pumps for mechanical exhaust cooling, bottom ash trough filling, and bottom ash crusher sealing. However, these pumps have lower efficiency due to lower flow and higher head [15]. Electrical parameters taken during the site measurement are tabulated in Table 5.9.

The detailed power-saving calculation is also given below.

Specifications of the new pump:
Flow = 55 m^3/h; head = 40 m; pump efficiency = 70%
Rating of new pump required = 11.0 kW
Estimated input power consumption by the new pump = 9.70 kW
The present average power consumption of one LP water pump = 15.8 kW

Table 5.9 Low-pressure water pump (AHP)

S.no.	*Parameters*	*HP water pump # 1 (fly ash operation)*	*HP water pump #2 (fly ash operation)*
A.	**Design parameters**		
1	Name of plant	Ash handling plant	
2	Make and type	Beacon Weir, SDC 200/200	
3	Rated capacity	600 m^3/h	
4	Speed (rpm)	1480	
5	Head (m)	80	
6	Pump efficiency	NA	
7	Motor rating (kW)	190kW	
B.	**Measured parameters**		
8	Voltage (kV)	6.69	6.67
9	Load current (ampere)	16.4	16.2
10	Power factor	0.89	0.9
11	Input power (kW)	169	168
12	Frequency (Hz)	49.2	49.1
13	Flow (m^3/h)	466	450
15	Suction /discharge pressure (kg/cm^2)	0.1/8.4	0.1/8.2
C.	**Performance calculations**		
16	Load (%)	84	84
17	Total head developed (m)	83	81
18	Computed motor efficiency (%)	94	94
19	Calculated pump output (kW)	105	99.3
20	Motor shaft power (kW)	158.6	158
21	Overall pump efficiency (%)	62	59
22	Computed motor efficiency (%)	94	94
23	Computed efficiency of the pump (%)	66	63
24	Design efficiency at duty point of similar rating pump (%)	70	70
25	Throttling (% open)	100	100

Power saving potential = 6.1 kW
Units saved = 48,312 kWh

Considering the 330 days and 24 hours of operation of one LP pump
Amount saved per year @ Rs. 2.0/kWh = Rs. 0.97 lakhs
The approximate investment required for an 11-kW new pump (2 no.) = Rs. 1.0 lakhs
Payback period = 12.4 months

5.9 ASH HANDLING PLANT LIGHTING LOAD (CHP)

The CHP Stage 2 uses lamps to illuminate various areas, measuring lighting burden both day and night [16]. Electrical parameters taken during the site measurement are tabulated in Tables 5.10 and 5.11.

Reducing the lighting supply voltage to 230 volts
The optimal lighting supply for lumens is 230 volts, as higher voltages increase lamp power consumption, reduce burning life, and increase failure rates [17].

Night-time

Total lighting load during night-time = 65.4 kW
Supply voltage measured = 237–254 volts
Power saving potential by maintaining 230 volts in the lighting feeder
(Calculated LDB-wise) = 5.75 kW
Units saved = 25,185 kWh
Considering 365 days and 12 hours of operation per day:

Daytime

Total lighting load during daytime = 19.1 kW
Supply voltage measured = 236–256 volts
Power saving potential by maintaining 230 volts in the lighting feeder [18,19].
(Calculated LDB-wise) = 1.15 kW
Units saved = 5,037 kWh
Considering 365 days and 12 hours of operation per day:
Total units saved per annum = 30,222 kWh
Amount saved per year @ Rs. 2.0/kWh = Rs. 0.60 lakh
Approximate investment required = Nil

Table 5.10 Lighting load analysis (night-time, CHP)

S. no.	*Description*	*Location*	*Rating (kVA)*	*Volt*	*Amp*	*P.F.*	*kW*	*kVA*	*Load (%)*	*Remarks*
1	LDB # 1	MCC Control	200	237	245	0.70	40.6	58.0	29	Higher voltage
2	LDB # 4	Room, Stage-2	100	254	59.1	0.60	9.01	15.0	15	Higher voltage
3	LDB # 2		200	244	113	0.57	15.8	27.7	14	Higher voltage
4	LDB # Stacker Reclaimer	Stack Reclaimer	10	234	37	0.60	5.2	8.72	87	Optimum voltage
5	LDB # 2	Wagon Tripler	100	231	162	0.82	30.7	37.5	37	Optimum voltage
6	LTG TR	TT-5	63.0	235	191	0.76	34.2	45.0	71	Optimum voltage

Table 5.11 Lighting load analysis (daytime, CHP)

S. no.	*Description*	*Location*	*Rating (kVA)*	*Volt*	*Amp*	*P.F.*	*kW*	*kVA*	*Load (%)*	*Remarks*
1	LDB # 1	MCC Control	200	237	82.6	0.8	15	20	10	Higher voltage
2	LDB # 4	Room, Stage-2	100	256	7.1	0.3	0.5	1.8	2	Higher voltage
3	LDB # 2		200	236	24.6	0.6	3.7	5.8	3	Higher voltage
4	LDB # Stacker Reclaimer	Stacker Reclaimer	10	234	0	0	0	0	0	Optimum voltage

5.10 CONCLUSION

The report suggests adjusting output loads to improve plant performance and reduce electrical energy waste. It recommends installing VFDs to regulate CEP speed, redesigning the water-cooling system, switching the interconnection control valve to automatic mode, and setting transformer tap positions to 230 volts. Implementing these recommendations could save around Rs. 91 lakhs annually, generate Rs. 142 lakhs in investments, and achieve an annual energy saving of 4.57 MUs.

REFERENCES

1. A. Boharb, A. Allouhi, R. Saidur, T. Kousksou, A. Jamil, "Energy conservation potential of an energy audit within the pulp and paper industry in Morocco," *Journal of Cleaner Production*, Vol.149, pp. 569–581, 15 April 2017, DOI:10.1016/j.jclepro.
2. D. Rathod, A. Tiwari, "Efficiency evaluation and optimization of a 210 MW thermal power plant: A comprehensive energy audit," *Journal of Thermal Engineering and Applications*, Vol.10, no.3, pp. 1–21,September–December 2023, doi:10.37591/JoTEA.
3. D. Rathod, N. Maithil, N. Khatri, "A case study referring to the energy audit in a captive thermal power plant," *Imperial Journal of Interdisciplinary Research (IJIR)*, Vol. 2, no.7, pp.966–974, 2016, ISSN: 2454-1362.
4. E. Andersson, M. Karlsson, P. Thollander, S. Paramonova, "Energy end-use and efficiency potentials among Swedish industrial small and medium-sized enterprises – a dataset analysis from the national energy audit program," *Renewable and Sustainable Energy Reviews*, Vol.93, pp.165–177, October 2018, doi:10.1016/j.rser.2018.05.037.
5. Z. Liu, H. Zhang, Y.-J. Zhang, T.-T. Zhu, "How does industrial policy affect the eco-efficiency of industrial sector? Evidence from China," *Applied Energy*, Vol. 272, pp.115–206, 2020.
6. D. Rathod, R. Khandare, A.K. Pandey, "Electrical energy audit (case study of tobacco industry)," *International Journal of Engineering and Applied Sciences*, Vol. 2, no.3, pp.9–18, 2013, ISSN: 2305-8269.
7. J. Rosenqvist, P. Thollander, P. Rohdin, M. Söderstrom, *Industrial Energy Auditing for Increased Sustainability – Methodology and Measurements*, InTech, January 2013, doi:10.5772/51717.
8. M. Iten, U. Fernandes, M.C. Oliveira, "Framework to assess eco-efficiency improvement: Case study of a meat production industry," *Energy Reports*, Vol.7, pp.7134–7148, November 2021, doi:10.1016/j.egyr.2021.09.120.
9. A. Tiwari, P. Sharma, "Preference-based grey theory model and its application in waste disposal selection: A case study," *Sadhana*, Vol.49, no.1,64, February2024, https://doi.org/10.1007/s12046-023-02413-8.
10. E. Sterling, C. Collett, S. Turner, C. Downing, "Commissioning to avoid indoor air quality problems," *ASHRAE, ATLANTA, GA(USA)*. Vol.67, p. 872, 1993.

11. V.Y. Anisimova, N.M. Tukavkin, E.K. Chirkunova, "Energy audit of industrial enterprises on the basis of energy and resource-saving evaluation," in: *E3S Web of Conferences*, vol. 91, p. 03003. EDP Sciences, 2019.
12. R.S. Chanda, "On energy auditing: An experience with a jute mill," *International Journal on Recent and Innovation Trends in Computing and Communication*, Vol.2, no.10, pp.3134–3137, October 2014.
13. A. Kluczek, P. Olszewski, "Energy audits in industrial processes," *Journal of Cleaner Production*, Vol.142, no.4, pp.3437–3453, November 2016, doi:10.1016/j.jclepro.2016.10.123.
14. M. Sharma, V.A. Koushik, "Energy audit: Case study of a wheel manufacturing industry," *International Journal of Engineering Research & Technology*. Vol.4, no.3, pp.15–23, 2015.
15. O. Sunday Olayinka, A.T. Oladele, "Energy audit of manufacturing and processing Industries in Nigeria: A case study of food processing industry and distillation and bottling company," *American Journal of Energy Research*, Vol.1, no.3, pp.36–44, January 2013, doi:10.12691/ajer-1-3-1.
16. M. Dongellini, C. Marinosci, G.L. Morini, "Energy audit of an industrial site: A case study," *Energy Procedia*,Vol.45, pp. 424–433, December 2014,doi:10.1016/j.egypro.2014.01.046.
17. V. Ghalandari, M.M. Majd, A. Golestanian, "Energy audit for pyro-processing unit of a new generation cement plant and feasibility study for recovering waste heat: A case study," *Energy*,Vol.173, pp.833–843, February 2019, doi:10.1016/j.energy.2019.02.102.
18. K.R. Ullah, M. Thirugnanasambandam, R. Saidur, K.A. Rahman, M.R. Kayser, "Analysis of energy use and energy savings: A case study of a condiment industry inIndia," *Energies*,Vol.14, no.16,4798, August 2021, https://doi.org/10.3390/en14164798.
19. N. Munguia, L. Velazquez, T. Bustamante, R. Perez, J. Winter, M. Will, B. Delakowitz, "Energy audit in the meat processing industry – a case study in Hermosillo, Sonora Mexico," *Journal of Environmental Protection*, Vol.7, no.1, pp.14–26, January (2016), doi:10.4236/jep.2016.71002.

Chapter 6

Application of AI in fault detection and grid optimization

Nagendra Singh, Amit Kumar Chandanan, Prachi Wadhawan, Igor Volynets, Anuprita Mishra, and Sanjeev Kumar Gupta

6.1 INTRODUCTION

The power system has a complex network, and its grid system has a wide infrastructure. The power system is interconnected with the different generating units in terms of the grid. Electrical power is the backbone of modern society. For the suitable operation of an electrical grid system, it is required to set up a reliable and efficient operation network. For reliable and stable operation of electrical power systems, it is required to adopt advanced techniques for fault detection and grid optimization [1].

For the protection and control of the abnormal operation of power system transmission, classical methods are used. Such classical methods sometimes operate the relays and circuit breakers for small changes in current and voltage. This is an unwanted interruption of the power supply. We need an advanced system that can understand the difference between fault conditions and small abnormal conditions [2].

Artificial intelligence (AI) handles such conditions and when faults arise in the system immediately, it gives a signal and operates the protective devices. It cannot open the network under small abnormal operating conditions. Artificial intelligence is one of the most recent and powerful tools that can be applied to monitor grid systems. It can help find and diagnose various faults in the grid system quickly. In such a way, it can improve the performance of electrical grids [3].

In this chapter, we will analyze the application of AI in fault detection and also find out how to optimize the performance of the grid system. This chapter also explores various technologies and methodologies that can be used in the real-time application of the power grid.

6.2 GENERAL FAULTS AND PROTECTION SYSTEM

The main objective of the electrical power system is to maintain stable operating conditions. If any abnormal operating conditions arise due to changes in load or faults in the system, clear them in the minimum amount

DOI: 10.1201/9781003516156-6

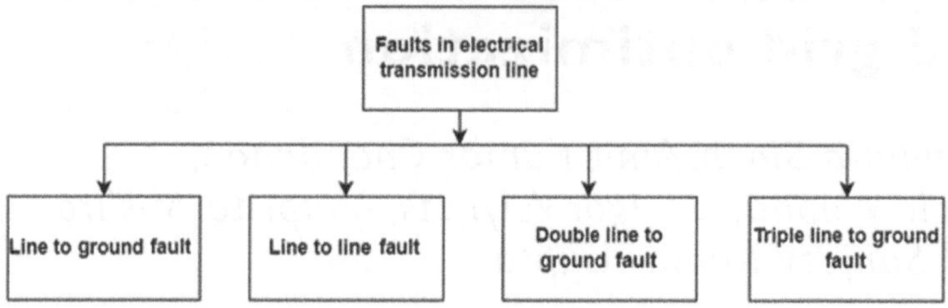

Figure 6.1 Types of faults in transmission line.

of time and maintain their stability. Electrical power system operation is affected by various causes, like environmental effects, equipment failure, physical accidents, and delays in control due to the old age of equipment. Due to all such kinds of effects, the voltage and current of the power system change suddenly, which is dangerous for electrical equipment and human beings [4].

Due to environmental accidents like the falling of a tree (wind effect) or lightning in rainy sessions, faults arise in the transmission line (line-ground fault, double line-to-ground fault, triple line-to-ground fault, and line-to-line faults). Figure 6.1 shows the types of faults in electrical transmission lines. When any kind of fault occurs in the transmission line, voltage decreases and current increases more than the safety limits. Overcurrent limits the most dangerous situation in the power system. Due to overcurrent, other parameters of the transmission line, like power factor, impedance, power limit, and frequency, also change, making the whole system unstable. Even this time, we are not able to operate the power system [5].

For the protection of such an abnormal operation in a power system, a switch gear and protection system are used. In this system, they used fuses and relays to identify abnormal conditions. The relay is connected to the circuit breaker; it gives a signal when an abnormal condition occurs. According to the signal of the relays, a circuit breaker is operated to break the affected lines from the networks [6].

Relays help in identifying the actual fault conditions, minimizing the duration of the operation of the power system in a fault condition, reducing the trouble time, and limiting the damage and outage time. A classical protection system is shown in Figure 6.2. The relay is connected to the transmission line using a correct transformer. A current transformer (CT) is used to read the flow of current from the transmission line. A circuit breaker is connected to the transmission line with a trip coil.

During abnormal conditions, the first CT reads the current value, and if it is greater than the safety limit, it sends a signal, which means an abnormal situation arose. In that situation, the lever of the trip coil is connected,

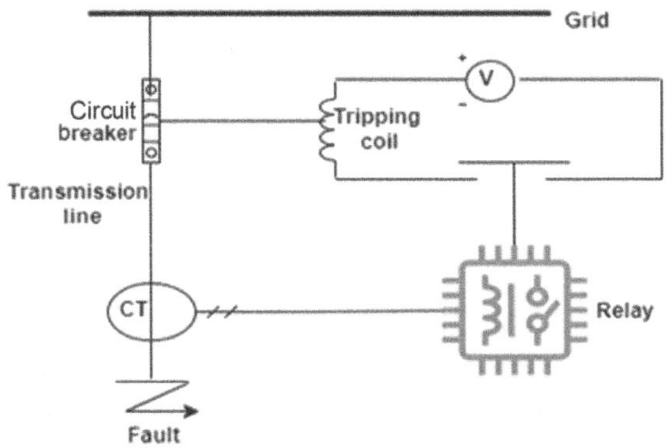

Figure 6.2 Traditional operation of relay and CB for the protection of faults.

which initializes the trip coil. Trip coil-initiated circuit breakers break the faulty section of the transmission line so that the remaining parts of the line and equipment remain under control and are safe [7].

6.2.1 Limitations of classical control system

The classical protection and control system has the following limitations or demerits:

- It required many instruments, like a CT, relay, circuit breaker, and trip coil.
- Such large instruments require a large space for installation.
- The cost of individual instruments is very high, so the overall cost also increased.
- The feature extension of such a system is not available.
- Relays and circuit breakers require regular maintenance.
- Due to aging, a relay does not operate in time; in that case, a big loss of network is possible.
- As it is a mechanically operated device, the speed of operation is also limited.

6.3 AI TECHNIQUES FOR FAULT DETECTION

Classical protection techniques have many disadvantages and limitations. Such limitations can be resolved by using the latest techniques in AI. AI techniques first read the historical data and, according to different situations, make the correct decisions. This is a great advantage of AI. Such features of the AI help to make the correct decision at the correct time [8].

In the transmission line, an abnormal situation arises for a few fractions of a second (not a fault condition). In that case, the relay also operates, and hence the circuit breaker opens the line and shuts down the power supply. AI detects the correct situation with the help of historical data and takes the correct action. In that case, an unnecessary shutdown of the network does not arise.

AI technology uses rules to represent the protection and circuit breaker actions in logical form and saves historical information in the knowledge base (KB). According to those rules and database information, it helps to identify and diagnose faults in the power network. When any type of fault occurs in the transmission line, the latter compares the fault with the rules in the KB and generates signals to operate the protective device.

The single line as well as the multi-network line use AI techniques for the identification and solving of abnormal conditions. AI has good computational and data processing capability to identify abnormal conditions very quickly, which is very useful in fault identification and diagnosis. AI has two categories: machine learning (ML) and deep learning (DL).

6.3.1 Machine learning

ML is a discipline of AI. It allows machines to learn automatically from past data and experience. According to available data, identify the correct patterns and make accurate predictions.

ML is one of the most exciting technologies ever encountered. As the name suggests, it gives computers something that makes them more like humans: the ability to learn. ML automates the computer's operation using algorithms or programming.ML algorithms learn from the data and, accordingly, make decisions. So ML is required to update the data continuously to make a perfect automation algorithm [9].

ML tries to understand the patterns of the past data and, with the help of an algorithm, gives a final analysis. The algorithm used the iterative process to find the best solution for the feed data. ML develops different types of models according to the given problems. As the number of samples increases, the performance of ML algorithms also improves accordingly because they learn from the sample data. Types of ML are shown in Figure 6.3.

6.3.1.1 Supervised learning

Supervised learning is a type of ML. In this type of learning approach, machines learn situations from supervised or known data sets. According to known data sets, it predicts situations and makes decisions.

In this learning, AI agents only find out the matching situations as given in the data set. Such a learning approach is divided into two categories:

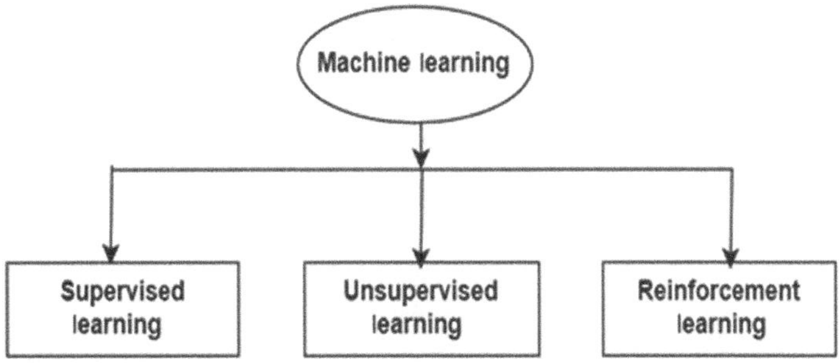

Figure 6.3 Types of ML.

- Classifications
- Regression

6.3.1.2 Unsupervised learning

Unsupervised learning refers to learning without any guidance. In this learning algorithm, unclassified data can be read. In this learning, AI agents learn from patterns in the data without corresponding output. In this learning process, two types of algorithms can be used:

- Clustering
- Association

6.3.1.3 Reinforcement learning

It is also a type of ML, where the working AI agent is first trained by specific useful commands, and according to training, it starts taking action. The agent uses this feedback to improve its performance.

- Positive reinforcement learning
- Negative reinforcement learning

All the types of ML methods mentioned above can be used for the supervision and control of electrical faults in power systems. Figure 6.4 shows the steps used by ML in fault detection.

6.3.2 Deep learning

DL, a branch of ML, develops algorithms using deep layers of neural networks, so it can work like a human without any supervision from

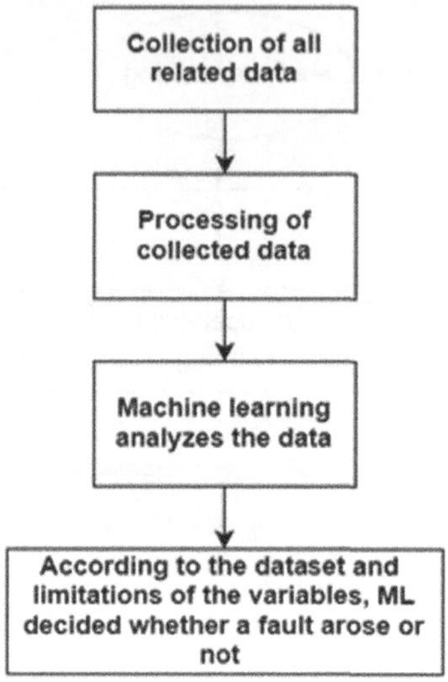

Figure 6.4 ML steps used for fault detection.

humans. DL develops an algorithm that can copy the behavior of the human brain. In this learning technique, supervised and unsupervised data are used for the development of agent models [10].

DL is also called deep neural networks because, in DL algorithms, the architecture of neural networks is used. DL has many applications, such as in self-driving cars, image recognition, automatic machine translation, and speech recognition. The main drawback of DL techniques is that they require a large amount of historical data as well as strong computing power. Figure 6.5 shows various layers of DL used in different applications [11].

6.3.2.1 Working of DL method

- DL algorithms are called so because they use deep neural networks.
- DL is divided into layers, with the first layer used as the input layer, the last layer used for the output, and many hidden layers also available for internal processes.
- The first layer receives the input and sends it to the hidden layer using neurons.
- Mathematical performance is done by hidden layers that pass the processed data to the last layer, called the output layer.

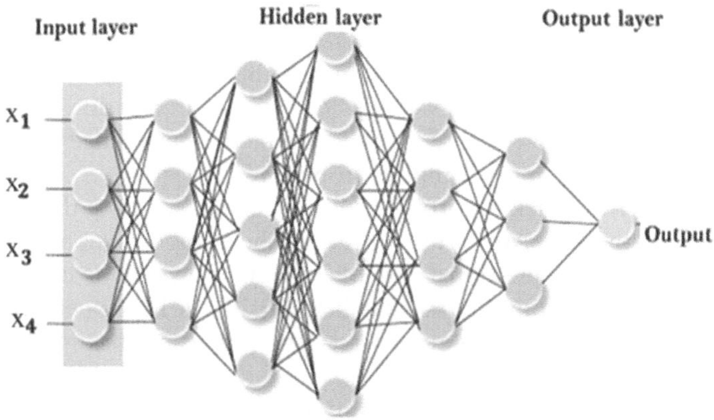

Figure 6.5 Deep learning network used for fault identification.

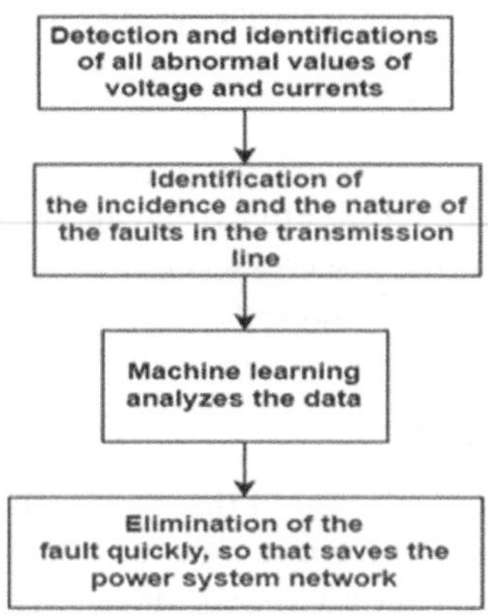

Figure 6.6 Deep learning techniques used to identify and remove the faults.

Fault diagnosis and prediction attempt to detect and localize the faults based on the collected sensor data and also predict their failures in advance, which can significantly help to take appropriate maintenance measures and avoid serious consequences for the power grid [12]. In recent years, DL methods have been widely used for fault detection and diagnosis due to their powerful feature representation capability and rapid evolution. Traditional intelligent fault diagnosis involves three different steps, namely signal detection, feature extraction, and fault classification, as shown in Figure 6.6.

6.3.3 Expert systems

The expert system (ES) was first successfully introduced as part of AI in 1970. The ES stores the information in the KB. Depending on the situation, ESs use reasoning and inference rules to extract information from the KB and take the correct action. With the help of the knowledge stored in the KB, an ES solves the most complex problems [13].

An ES solves the trickiest problems like an expert by using the knowledge contained in the KB. An ES uses facts and heuristics to support decision-making on difficult questions, just like a human expert. It can solve any complicated problem in a particular field because it has expert knowledge in that field. Error detection and correction are the specific areas for which ESs are developed [14].

Figure 6.7 shows a block diagram of an ES (components). The KB of an ES contains the expert knowledge with which it works. The more knowledge is stored in the KB, the more powerful the system is. The Google search box, which suggests the correct spelling after a user has entered an incorrect word, is one of the best-known examples of an ES [15].

The basic components used in an ES are discussed below.

6.3.3.1 User interface

It displays the output to the user after receiving the response from the inference engine. In other words, it serves as an interface for non-expert users to interact with ESs and obtain solutions.

6.3.3.2 Rule-based engine

As it is the primary processing unit of the system, the inference engine is also referred to as the brain of the ES. It uses the KB and the inference rules to reach a conclusion or extract additional data. It helps to find an error-free answer to the questions posed by the user. The system extracts knowledge from the KB using an inference engine [16].

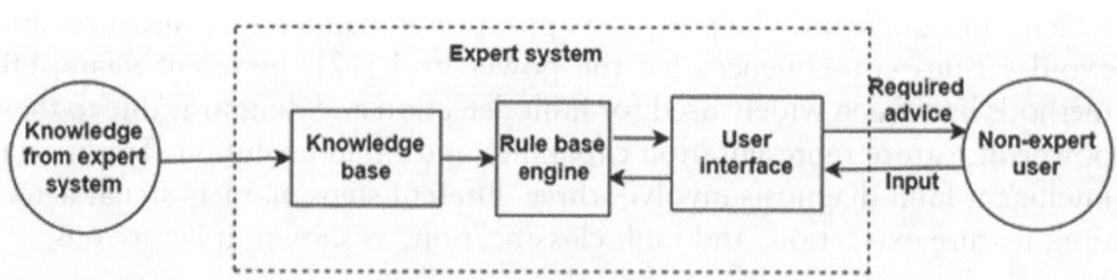

Figure 6.7 Various components of expert system.

The following two types of inference engines are used:

- *Deterministic inference engine:* In this inference system, the conclusions are based on facts and rules. The conclusion is considered to be true.
- *Probabilistic inference engine:* This inference system contains uncertainty, and on the basis of probability, the conclusions are drawn.

The inference engine drives the solution using two types of modes:

- *Forward chaining:* In this model, the system solution starts with the known facts and rules and moves to the final goal.
- *Backward chaining:* This model system starts from the goal to the initial point. It is used to fine-tune the facts and rules used by the system to find the goal.

6.3.3.3 Knowledge base

It is a form of memory that stores the knowledge and information acquired by various experts in a particular field. KB is a large knowledge store. The more knowledge experts store in the KB, the better the results an ES delivers [17]. It is like a database that contains a lot of information and rules for a specific area. Figure 6.8 shows the block diagram of an ES used for fault detection and diagnosis.

6.3.3.4 Characteristics of ES

- *High performance:* ESs are highly effective at solving complicated problems in a particular domain with a high degree of efficiency and accuracy.

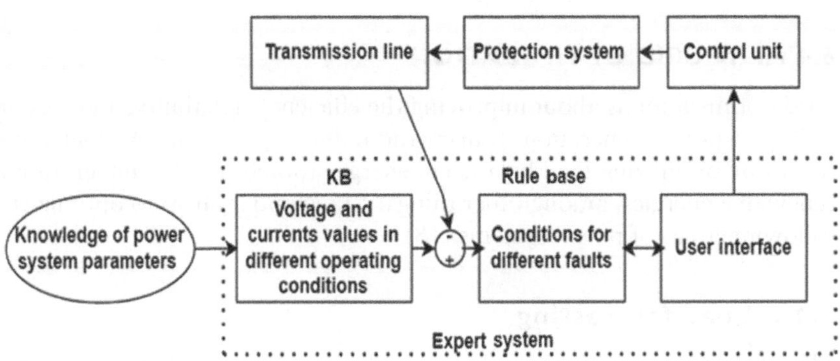

Figure 6.8 Expert system arrangements for fault detection.

- *Understandable:* ESs answer in easily understood human language after receiving input in basic human language.
- *Reliable:* It provides the most accurate results based on the information stored in the KB.
- *Highly responsive:* It provides quick answers to even complex questions.

6.3.3.5 Advantages and limitations of ES

6.3.3.5.1 Advantages of ES

- Memory limitations are not applicable.
- Highly reproducible.
- They can be used in risky places where the presence of humans is not safe.
- They give an accurate response if the KB contains the right information.
- The performance of the systems is not affected by emotions, tension, or fatigue.
- It provides a very fast response to a specific request.

6.3.3.5.2 Limitations of ES

- The response of the ES depends on the information that is fed into the KB. So, if the information is incorrect, the answer will also be incorrect.
- It cannot make decisions like a human being in various situations.
- The development and maintenance of an ES are very expensive.
- When developing an ES for a specific area, it is very difficult to acquire knowledge.
- Each domain requires a specific ES.
- The KB system must be constantly updated.

6.4 AI IN GRID OPTIMIZATION

Grid optimization is about improving the efficiency, reliability, and sustainability of power generation, transmission, and distribution. AI contributes to this by optimizing load balancing, energy storage, and the integration of renewable energies, among other things. AI techniques help to optimize the following areas of power systems [18].

6.4.1 Load forecasting

ML algorithms are better able to track seemingly random trends than forecasting methods such as curve fitting using numerical techniques, which lead to inaccurate results in the complicated, multivariable, and multidimensional estimation problem of load forecasting [19].

Knowing the different features that influence electricity consumption is the first step in creating an ML model for load forecasting. It depends on a number of factors, including the time of day, historical patterns of electricity demand, humidity, weather, and electricity costs. The stability of the electricity grid depends on accurate load forecasts.

AI models, especially DL algorithms such as long short-term memory (LSTM), can predict short- and long-term electricity demand with high precision, enabling better planning and resource allocation.

6.4.2 Optimal power flow

AI models, especially DL algorithms such as LSTM, can predict short- and long-term electricity demand with high precision, enabling better planning and resource allocation. AI has played an important role in various areas of electricity systems. With the help of AI, we can plan the expansion, forecasting, and stability of electricity systems. AI has been developed with advanced technologies and is therefore able to support the optimization of power flow in electricity systems [20].

In general, economic load flow optimization is used to determine the best operating costs for thermal power plants. It also helps to find the minimum emissions in the environment. Using various optimization techniques such as particle swarm optimization (PSO), New PSO, FUZZY system, etc., it is possible to find the optimal solution for both a single objective and a multicriteria economic load balancing.

6.4.3 Energy storage management

The demand for electrical energy changes throughout the day. AI forecasts the demand for electricity and the available supply of electricity and optimizes the use of electricity from the storage system during the peak load hour. AI helps determine the optimal time to charge the battery system and utilize the battery power. It helps to increase the efficiency and service life of the storage batteries [21].

6.4.4 Renewable energy integration

Renewable energy sources are used to generate electricity, but they face many problems. Solar panels only generate electricity during the day, as sunlight is only available during the day. Wind turbines also have problems with irregular wind conditions. In all these situations, AI helps in accurately predicting the energy generated by renewable energy sources and thus connecting them to the power grid. AI helps in the development of hybrid power generation systems [6].

6.5 CHALLENGES AND FUTURE DIRECTIONS IN AI

In energy systems, AI is used to improve performance, increase the utility of energy, and address various problems associated with its operation. Below we list the various challenges and future developments in AI.

Challenges for AI in power system

- *Data quality and quantity:* For the best performance of AI, its model requires a large quantity of power system data. The data should be correct and be of good quality.
- *Integration with legacy systems:* Most power systems operate with old infrastructure, so the design of an AI model is very difficult and also very costly.
- *Cyber security:* Cyber-attacks are very common, so it is essential to design a secure AI model for power systems. Hackers can steal very sensitive information about power system operation and management.
- *Regulatory and ethical considerations:* Power systems operate under many rules and regulations given by the government, so when developing an AI model for a power system, it should not violate these rules and regulations.

Future developments required in AI

1. *Advanced ML models:* Continued development of more sophisticated ML models, including hybrid approaches combining various AI techniques, will enhance fault detection and grid optimization capabilities.
2. *Edge computing:* Implementing AI at the edge, close to where data is generated, will reduce latency and improve real-time decision-making.
3. *AI and IoT integration:* The synergy between AI and the Internet of Things (IoT) will lead to more interconnected and intelligent grid systems, capable of self-healing and autonomous operation.
4. *Sustainability focus:* AI will play a critical role in achieving sustainability goals by optimizing energy usage, integrating more renewable sources, and reducing carbon footprints.

6.6 CONCLUSIONS

AI plays an important role in fault detection and grid optimization. AI can handle the large, complex data of power systems and make accurate decisions in fault conditions. AI enhances reliability in operation, improves system efficiency, provides stability in power system operation, and improves the life span of the equipment. It can also help to interconnect renewable energy sources into the grid system. AI models help operate the power system as a hybrid plant.

At the same time, AI is also used to optimize the grid operation. As it has many advanced techniques, AI can easily optimize large amounts of data quickly.

REFERENCES

1. Rohini, G., Gnana Kousalya, C., Singh, N., Patil, V.R. Autonomous forecasting of traffic in cellular networks based on long-short term memory recurrent neural network. *Cybernetics and Systems*, 56(1), 33–45, 2023. https://doi.org/10.1080/01969722.2023.2166245
2. Almasoudi, F.M. Enhancing power grid resilience through real-time fault detection and remediation using advanced hybrid machine learning models. *Sustainability*,15(10),8348, 2023. https://doi.org/10.3390/su15108348
3. Abubakar, A., Jibril, M.M., Almeida, C.F.M., Gemignani,M.,Yahya, M.N., Abba, S.I. A novel hybrid optimization approach for fault detection in photovoltaic arrays and inverters using AI and statistical learning techniques: A focus on sustainable environment. *Processes*, 11(9), 2549, 2023. https://doi.org/10.3390/pr11092549
4. Omitaomu, O.A.,Niu, H. Artificial intelligence techniques in smart grid: A survey. *Smart Cities*,4, 548–568, 2021. https://doi.org/10.3390/smartcities4020029
5. Zulu, M.L.T., Carpanen, R.P., Tiako, R. A comprehensive review: Study of artificial intelligence optimization technique applications in a hybrid microgrid at times of fault outbreaks. *Energies*,16(4), 1786, 2023. https://doi.org/10.3390/en16041786
6. Behara, R.K., Saha, A.K. Artificial intelligence methodologies in smart grid-integrated doubly fed induction generator design optimization and reliability assessment: A review. *Energies*,15(19), 7164, 2022. https://doi.org/10.3390/en15197164
7. Shi, Z., Yao, W. Artificial intelligence techniques for stability analysis and control in smart grids: Methodologies, applications, challenges and future directions. *Applied Energy*, 278, 1–25, 2020. https://doi.org/10.1016/j.apenergy.2020.115733
8. Gao, D.W. Application of AI techniques in monitoring and operation of power systems. *Frontiers in Energy*, 13(1),71–85, 2019. https://doi.org/10.1007/s11708-018-0589-4
9. Khadse, C., Patharkar, A.A. Electromagnetic field and artificial intelligence-based fault detection and classification system for the transmission lines in smart grid. *Energy Sources, Part A: Recovery, Utilization, and Environmental Effects*, 1–15, 2021. https://doi.org/10.1080/15567036.2021.1948637
10. Mhdawi, A.L.,Raweshidy, A.L. A smart optimization of fault diagnosis in electrical grid using distributed software-defined IoT system. *IEEE Systems Journal*, 14(2), June2020. doi:10.1109/JSYST.2019.2921867

11. Zhang, F., Saeed, N., Sadeghian, P. Deep learning in fault detection and diagnosis of building HVAC systems: A systematic review with meta analysis. *Energy and AI*, 12, 1–28,100235, 2023. https://doi.org/10.1016/j.egyai.2023.100235.
12. Singh, N., Chakrabarti, T., Chakrabarti, P., Margala, M., Gupta, A., Praveen, S.P., Krishnan, S.B., Unhelkar, B. Novel heuristic optimization technique to solve economic load dispatch and economic emission load dispatch problems. *Electronics*, 12(13), 2921, 2023. https://doi.org/10.3390/electronics12132921.
13. Rohini, G., Gnana Kousalya, C., Singh, N., Patil, V.R. Autonomous forecasting of traffic in cellular networks based on long-short term memory recurrent neural network. *Cybernetics and Systems*, 56(1), 24 January 2023. doi.org/10.1080/01969722.2023.2166245
14. Veena, R.S.,Khambra, G.,Sayal, R.,Singh, N., Bala Dhandayuthapani, V., Sar, S.K. An energy efficient IPv6 packet delivery outline for industrial IoT over G.9959 protocol based wireless sensor network (WSN). *SN Computer Science*,5, 813,2024. https://doi.org/10.1007/s42979-024-03195-4
15. Singh, N., Kumar Prajapati, G. Wireless sensor network security problems –a review. *International Journal of Emerging Technology and Advanced Engineering*, 8(1), January 2018.
16. Singh, N., Kumar, Y. Multi objective economic load dispatch problem solved by new PSO. *Advances in Electrical Engineering*, 2015, 2356–6655, 2015.
17. Singh, N., Tamrakar, S., Mewada, A., & Gupta, S.K.(eds.). *Artificial Intelligence Techniques in Power Systems Operations and Analysis* (1st ed.). 2023. Auerbach Publications. https://doi.org/10.1201/9781003301820
18. Singh, N. Application of Fiber Optics for the Protection and Control of Power Systems. In: Saha, A., Biswas, A., Ghosh, K., Mukhopadhyay, N. (eds.) *Optical to Terahertz Engineering. Progress in Optical Science and Photonics*, vol. 23, 2023. Springer, Singapore. https://doi.org/10.1007/978-981-99-0228-6_7
19. Sondhiya, V., Dwivedi, K., Kulsum Almas, S., Singh, N. Energy Management of Nanogrid through Flair of Deep Learning from IoT Environments. In: *Artificial Intelligence Techniques in Power Systems Operations and Analysis*, Taylor& Francis, July 2023.
20. Singh, N., Mishra, A., Gupta, M., Dwivedi, K., Swarnkar, M. Implementation of Microgrid for Rural Electrification to Cater on Side Load Demand. In: *2023 3rd International Conference on Energy, Power and Electrical Engineering (EPEE)*, Wuhan, China, 2023, pp. 1058-1061, doi: 10.1109/EPEE59859.2023.10351810.
21. Singh, N. Meta Heuristic Techniques Used to Improve the Performance of Static Converters. In: *International Conference on Innovative Development in Virtual and Rapid Prototyping (IDVRP-2022)*. AIP Conference Proceedings 2771, 010001, 2023.

Chapter 7

Agricultural sustainability and the use of renewable energy sources

Sateesh Kourav, Kirti Verma, M. Sundararajan, and Bhargavi Jha

7.1 INTRODUCTION

Integrating renewable energy into agricultural practices is a game-changing solution for tackling the pressing issues of agricultural sustainability and energy security. Given the pressing issues of a growing global population and the escalating effects of climate change, it becomes crucial to devise creative solutions that boost agricultural productivity, mitigate environmental harm, and decrease dependence on limited fossil fuel resources. Renewable energy sources like solar, wind, biomass, and hydroelectric power provide a sustainable alternative to traditional fossil fuels [1]. They offer clean, abundant, and locally available energy for agricultural operations. Through the utilization of renewable energy, farmers have the opportunity to decrease greenhouse gas emissions, decrease energy expenses, and strengthen the durability of their operations against climate-related hazards. This introduction provides an overview of various aspects of integrating renewable energy into agriculture, such as its applications, advantages, obstacles, and potential for the future. Through an analysis of the impact of renewable energy on agricultural innovation, sustainable practices, and rural development, we can understand how the integration of renewable energy can support a stronger, fairer, and more environmentally friendly agricultural sector. In this discussion, we will explore different aspects of integrating renewable energy in agriculture [2].

This includes on-farm energy production, integrated renewable energy systems, policy frameworks, technological innovations, and successful case studies of renewable energy adoption in agricultural settings. Through the exploration of the potential connections between renewable energy and agriculture, we can discover fresh possibilities for promoting sustainable development goals, improving food security, and reducing the effects of climate change. Ultimately, the incorporation of renewable energy into agriculture offers great potential in promoting agricultural sustainability, achieving energy independence, and fostering rural development. Through the utilization of renewable energy, we have the ability to create a more robust,

DOI: 10.1201/9781003516156-7

effective, and environmentally aware agricultural industry that caters to the requirements of both current and future generations. With a focus on collaboration and creative problem-solving, we can pave the way for a more environmentally friendly and sustainable future in agriculture and beyond [3].

The passage emphasizes the pressing requirement to shift towards sustainable and environmentally friendly energy solutions, especially within the built environment, in light of the limited availability of conventional energy sources and the increasing worries about climate change. Given the impact of buildings on greenhouse gas emissions, there is an increasing focus on integrating renewable energy technologies into contemporary building design and construction. Given the rise in energy demand in the building sector due to urbanization and population growth, it is crucial to address emissions and shift towards net-zero energy buildings. The targets set by the Paris Climate Agreement highlight the urgent need to reduce emissions from buildings by 50% by 2030 and to achieve net-zero lifecycle emissions for all buildings by 2050. It is crucial to the reduction of carbon emissions from buildings, given that this sector currently contributes significantly to global energy consumption and greenhouse gas emissions [4].

Through a focus on sustainable practices and the incorporation of renewable energy technologies into building design and operations, stakeholders can make significant strides in addressing climate change and advancing global sustainability objectives. This passage highlights the crucial importance of incorporating renewable energy into the built environment. It not only helps to reduce emissions but also promotes energy efficiency, resilience, and sustainability in the face of climate challenges. In the global race to meet climate targets, the shift towards renewable energy in buildings is becoming increasingly important in our collective push for a carbon-neutral future. According to the International Energy Agency (IEA) and other sources, the building sector is estimated to account for approximately 40% of global energy consumption and about 37% of total greenhouse gas emissions worldwide. This encompasses emissions from energy consumption for heating, cooling, lighting, and other operational requirements in residential, commercial, and institutional buildings [5].

The specific distribution of CO_2 emissions from the building sector can vary across different regions, countries, and years. This variation is influenced by various factors, including the implementation of energy efficiency measures, the adoption of renewable energy sources, population growth, urbanization rates, and economic activities [6]. To obtain the most precise and current data on global building sector energy-related CO_2 emissions in 2021, it would be advisable to refer to reports and databases provided by esteemed organizations like the IEA, the United Nations Environment Programme (UNEP), or the World Green Building Council (GBC). These organizations frequently release in-depth analyses and statistics on global CO_2 emissions from the building sector. Integrating renewable energy sources into agriculture is vital for ensuring sustainable farming practices. Here's how various renewable energy sources can be applied in agriculture.

7.1.1 Solar energy

Solar power can be harnessed through photovoltaic (PV) panels to generate electricity for various farming operations. This energy can power irrigation systems, greenhouse lighting, electric fences, and machinery. Solar panels can be mounted on farm buildings or installed on unused land, minimizing their impact on agricultural activities [7].

7.1.2 Wind energy

Wind turbines can be installed on farms to generate electricity from the wind's kinetic energy. Wind power can supplement or replace traditional grid electricity, reducing energy costs and greenhouse gas emissions. Wind energy is particularly advantageous in regions with consistent wind patterns, such as coastal areas or open plains.

7.1.3 Hydropower

Small-scale hydropower systems can utilize flowing water from streams or rivers on or near farms to generate electricity. These systems offer a reliable source of renewable energy, particularly in areas with abundant water resources. Hydropower can power farm operations and reduce reliance on fossil fuels [8].

7.1.4 Biomass energy

Agricultural waste, such as crop residues, animal manure, and food processing byproducts, can be converted into bio-fuels through processes like anaerobic digestion or biomass gasification. Bio-fuels can generate heat, electricity, or power vehicles, offering a sustainable alternative to fossil fuels.

7.1.5 Geothermal energy

Geothermal heat pumps can be used for heating and cooling in greenhouses, livestock buildings, and other agricultural facilities. By harnessing the Earth's natural heat, geothermal energy systems significantly decrease energy consumption and minimize greenhouse gas emissions compared to traditional heating and cooling methods.

7.1.6 Biogas

Biogas production involves the anaerobic digestion of organic materials such as animal manure, crop residues, and food waste to produce a methane-rich gas. This biogas is a valuable resource that can be utilized as a clean and sustainable fuel for various purposes such as heating, cooking, and electricity generation on farms. By harnessing this renewable energy source, we can effectively reduce greenhouse gas emissions and contribute to a more sustainable future. Through the integration of renewable energy sources into agriculture, farmers have the opportunity to minimize their impact on the environment, decrease their dependence on non-renewable resources, and play a crucial role in addressing climate change [7].

The analysis highlights the intricate relationship between global energy demand, climate objectives, and the role of fossil fuels in meeting international climate targets. Although it may not be essential to completely eliminate fossil fuels to restrict global warming to 1.5°C or 2°C by 2100, substantial reductions in their usage are crucial. Scenarios indicate that fossil fuel use will decrease but remain significant until at least mid-century, emphasizing the need for a shift towards cleaner energy sources. This includes growing renewable energy like wind and solar and implementing carbon removal technologies such as direct air capture (DAC), bioenergy with carbon capture and storage (CCS), and nature-based solutions [9]. Relying solely on carbon removal technologies presents challenges and uncertainties, necessitating strong monitoring, reporting, and verification mechanisms. Ethics, social, and environmental concerns also need to be addressed. Achieving climate targets requires a comprehensive strategy integrating fossil fuel reduction with investments in renewable energy, energy efficiency, and carbon removal technologies. By embracing a diverse portfolio of solutions and prioritizing innovation, collaboration, and policy support, the global community can work towards a sustainable and resilient energy future while mitigating climate change impacts [10].

7.2 APPLICATIONS OF RENEWABLE ENERGY INTEGRATION FOR AGRICULTURE

Integrating renewable energy into agriculture presents a wide range of applications that can greatly improve sustainability, efficiency, and resilience in the sector [9]. Here are two important applications. Generating on-farm energy is a key use of renewable energy in agriculture. It helps meet the electricity and heating requirements of agricultural operations. This can be accomplished through a range of renewable energy technologies, such as solar PV systems, which are a great option for generating electricity from sunlight. They can be installed on farm buildings, barns, or unused land. This electricity has the potential to fuel farm equipment, irrigation pumps, lighting, and various other electrical appliances, thereby decreasing dependence on grid electricity and cutting down energy expenses.

7.2.1 Integrated renewable energy systems

An interesting application of renewable energy integration in agriculture is the development of integrated renewable energy systems that combine various renewable energy technologies to fulfil the diverse energy requirements of agricultural operations [10]. By leveraging the complementary characteristics of different renewable energy sources, integrated systems have the potential to enhance energy reliability, efficiency, and resilience.

7.2.2 Hybrid solar-wind systems

Combining solar PV panels with wind turbines allows for the capture of energy from both sunlight and wind. By utilizing a hybrid approach, energy production can be optimized regardless of the time of day or weather conditions. This ensures a dependable and steady energy supply for agricultural operations.

7.2.3 Biomass

Integrated systems that combine solar PV panels, biomass energy generation, and battery storage can offer a dependable and self-sustaining energy supply for off-grid or remote agricultural sites. During the day, electricity is generated by solar panels, with backup power from biomass energy and battery storage systems during times of low solar availability.

7.2.4 Renewable energy

Microgrids are designed to integrate various renewable energy sources, energy storage systems, and smart grid technologies. These systems create self-contained energy networks that cater to agricultural communities or clusters of farms. Microgrids offer a range of benefits, including optimized energy distribution, efficient energy demand management, and improved resilience to grid outages. They are an excellent decentralized and sustainable energy solution for rural areas [11].

7.3 ROLE OF THE RENEWABLE ENERGY INTEGRATION FOR AGRICULTURE

Renewable energy is critical for addressing significant global concerns such as climate change, energy security, and sustainable development. Its significance stretches across a variety of sectors and is particularly important for addressing these issues. Here are three significant functions that renewable energy may provide. Renewable energy is an essential component in the fight against climate change asit helps to cut down on emissions of greenhouse gases. Solar, wind, hydro, and biomass are examples of renewable energy sources that have the particular benefit of producing very little to no greenhouse emissions during operation [12].

Renewable energy plays a crucial role in bolstering energy security by diversifying the energy mix and decreasing reliance on limited fossil fuel resources. Renewable energy sources have distinct advantages over fossil fuels. They are abundant, widely distributed, and readily available, unlike fossil fuels which are prone to price fluctuations and geopolitical uncertainties [13].

Renewable energy plays a crucial role in promoting sustainable development as it supports economic growth, social equity, and environmental protection. Renewable energy projects have the potential to generate employment opportunities, boost local economies, and attract investments, especially in rural and underserved areas. In addition, the deployment of renewable energy has a positive impact on public health as it helps to decrease air and water pollution that is linked to the burning of fossil fuels [14].

Renewable energy integration is essential for mitigating the environmental impacts of agricultural activities, contributing to environmental sustainability, and climate mitigation. Renewable energy sources like solar and wind have a significant advantage over fossil fuels [15]. They generate very few greenhouse gas emissions while in operation, which means they can greatly reduce the carbon footprint of agricultural operations.

Climate change presents major obstacles to agricultural productivity, as extreme weather events, water scarcity, and changing growing seasons continue to affect crop yields and farm incomes. Renewable energy integration improves the resilience of agricultural systems by offering a decentralized and diversified energy source that is more resistant to disruptions.

Renewable energy technologies have the potential to bring electricity to remote and off-grid areas, making a positive impact on energy access and improving people's lives. By advocating for the adoption of renewable energy, nations can successfully attain the triple bottom line of sustainable development, encompassing economic prosperity, social equity, and environmental sustainability.

7.4 CHALLENGES AND LIMITATIONS

The incorporation of sustainable energy sources into agricultural practices has garnered considerable interest in recent years, driven by the urgency of achieving sustainable development and the necessity of reducing the effects of climate change. This study provides a thorough examination of the current status of research and practices concerning the incorporation of renewable energy technology in agriculture. It specifically highlights the potential advantages, difficulties, and prospects associated with this integration. The research explores the positive environmental and economic outcomes that come with implementing renewable energy sources in agriculture [11]. These include a decrease in greenhouse gas emissions, cost savings on energy, and improved ability to withstand the impacts of climate change. The study examines numerous renewable energy technologies that can be used in agricultural settings, including solar PV systems, wind turbines, biomass energy, and hydroelectric power. It emphasizes how these technologies are well-suited for various agricultural settings and geographical regions.

The article investigates the obstacles and impediments that are impeding the extensive use of renewable energy in agriculture, such as exorbitant initial expenses, technological limitations, and regulatory obstacles. The text discusses strategies for addressing these issues, such as offering financial incentives, implementing technological advances, and enacting legislative interventions to encourage the incorporation of renewable energy in agricultural activities [12]. The possible synergies between the generation of renewable energy and agricultural practices include the utilization of solar panels as shading structures for crops, wind turbines on agricultural land, and the creation of biogas from agricultural waste. The integration of these technologies not only aids in the sustainable production of energy but also improves the overall efficiency and productivity of agricultural operations [13].

The current research highlights the significance of comprehensive strategies for sustainability in agriculture. These strategies involve incorporating renewable energy sources along with other sustainable practices, including organic farming, water conservation, and biodiversity conservation. The statement emphasizes the importance of ongoing research, innovation, and policy assistance in fully harnessing the benefits of incorporating renewable energy into sustainable agriculture [14]. This integration is crucial for solving the issues of food security and mitigating the impacts of climate change. The integration of renewable energy into agricultural operations holds immense potential for promoting sustainability, resilience, and energy autonomy. However, several challenges must be effectively tackled to ensure successful implementation and widespread adoption [15]. This article identifies and explores four significant challenges facing the integration of renewable energy in agriculture, offering strategic insights to address them.

7.4.1 Variability

Renewable energy sources such as solar and wind exhibit inherent intermittency and variability, posing challenges for ensuring a reliable energy supply for agricultural activities. Strategic solutions include the deployment of energy storage systems to mitigate fluctuations, smart grid technologies for demand-response management, and the integration of complementary renewable energy sources to diversify the energy mix and enhance reliability.

7.4.2 Financial barriers

High upfront costs and limited access to financing often deter farmers from investing in renewable energy technologies. To overcome this challenge, innovative financing mechanisms tailored to the agricultural sector, such as leasing arrangements, performance-based incentives, and low-interest loans, can be introduced. Additionally, capacity-building initiatives and technical assistance programmes can help farmers navigate financial complexities and make informed investment decisions.

7.4.3 Regulatory and policy constraints

Inconsistent or inadequate regulatory frameworks, permitting processes, and land-use policies create several barriers to the deployment of renewable energy systems on agricultural land. Policy reforms aimed at streamlining permitting procedures, providing incentives for renewable energy adoption, and promoting agricultural energy production can facilitate smoother integration. Collaborative efforts between government agencies, agricultural stakeholders, and renewable energy industry players are essential to drive policy changes that support renewable energy deployment in agriculture.

7.4.4 Technological compatibility and adaptation

Integrating renewable energy technologies into existing agricultural infrastructure and practices requires careful consideration of technological compatibility and adaptation. Customized solutions tailored to the specific needs and constraints of agricultural operations, such as solar PV panels integrated into crop shading structures or wind turbines installed on marginal lands, can maximize synergies and minimize disruptions. Research and development efforts focused on developing innovative, cost-effective, and user-friendly renewable energy solutions for agricultural applications are crucial to overcoming technological barriers.

7.5 FUTURE DIRECTIONS AND RESEARCH OPPORTUNITIES

As the global agricultural sector seeks to transition toward sustainability and resilience in the face of climate change and energy challenges, the integration of renewable energy offers promising avenues for innovation and transformation. This article explores future directions and research opportunities in renewable energy integration for agriculture, aiming to identify strategies and technologies that can drive progress toward sustainable agricultural practices.

7.5.1 Technological advancements and innovations

Future research efforts should focus on advancing renewable energy technologies tailored to the specific needs and constraints of agricultural operations. This includes the development of efficient and cost-effective solar PV systems, wind turbines optimized for on-farm use, and innovative biomass energy solutions utilizing agricultural residues. Additionally, advancements in energy storage technologies, smart grid systems, and energy management platforms will be crucial for enhancing the reliability and resilience of renewable energy integration in agriculture.

7.5.2 Integration of renewable energy with precision agriculture

The integration of renewable energy with precision agriculture techniques holds great potential for optimizing resource use, enhancing productivity, and minimizing environmental impacts. Future research should explore synergies between renewable energy production and precision agriculture practices, such as the use of solar-powered sensors for real-time monitoring and decision-making, automated irrigation systems powered by renewable energy, and drone-based technologies for crop management and surveillance.

7.5.3 Climate-resilient agriculture and energy systems

Climate change poses significant challenges to both agriculture and energy systems, necessitating the development of climate-resilient solutions. Future research should focus on integrating renewable energy technologies with climate-smart agricultural practices to enhance resilience to extreme weather events, water scarcity, and other climate-related risks. This includes the adoption of agro-forestry systems combining renewable energy production with tree-based farming practices, the development of renewable energy-powered desalination technologies for irrigation in arid regions, and the implementation of microgrid systems to ensure energy security in remote agricultural areas vulnerable to climate impacts.

7.5.4 Policy support and institutional frameworks

Effective policy support and institutional frameworks are essential for accelerating the adoption of renewable energy in agriculture. Future research should explore policy mechanisms, regulatory frameworks, and incentive structures that incentivize renewable energy integration while addressing barriers and promoting equitable access for smallholder farmers and rural communities. This includes the development of feed-in tariffs, tax incentives, and subsidies for renewable energy deployment in agriculture, as well as capacity-building initiatives and knowledge-sharing platforms to empower farmers and stakeholders.

7.5.5 Cross-disciplinary collaboration and knowledge exchange

Collaboration between diverse stakeholders, including researchers, policymakers, farmers, industry actors, and civil society organizations, is critical for driving innovation and knowledge exchange in renewable energy integration for agriculture. Future research should prioritize cross-disciplinary collaboration, fostering partnerships that bridge the gap between agricultural and energy sectors, integrate local knowledge and expertise, and promote participatory approaches to technology development and decision-making.

7.6 CONCLUSION AND FUTURE SCOPE

Integrating renewable energy into agricultural practices presents a crucial chance to promote sustainability, resilience, and innovation within the agricultural sector. Throughout this discussion, we have delved into the numerous advantages, obstacles, and future prospects related to the integration of renewable energy in agriculture. Renewable energy technologies, such as solar PV systems, wind turbines, biomass energy, and hydroelectric power, provide a way to decrease greenhouse gas emissions, improve energy efficiency, and address the effects of climate change on agricultural productivity.

7.6.1 Precision agriculture integration

The integration of renewable energy with precision agriculture techniques represents a frontier of opportunity for optimizing resource management, enhancing productivity, and minimizing environmental impacts. Future research and development efforts will focus on leveraging renewable energy-powered sensors, drones, and automated systems to enable real-time monitoring, decision-making, and management of agricultural processes. By integrating renewable energy with precision agriculture, farmers can achieve greater efficiency, precision, and sustainability in their operations while reducing inputs and mitigating environmental impacts.

7.6.2 Climate-resilient solutions

Addressing the impacts of climate change on agricultural productivity and food security requires the implementation of resilient solutions that combine renewable energy with adaptive agricultural practices. In the future, our research will delve into the connections between renewable energy production and climate-smart agriculture techniques. This includes exploring agro-forestry, conservation agriculture, and water management strategies.

7.6.3 Policy support and market

The future of renewable energy integration in agriculture will be shaped by supportive policy frameworks, market incentives, and regulatory mechanisms that incentivize investment, innovation, and adoption. Governments, international organizations, and agricultural stakeholders will collaborate to develop policies that promote renewable energy deployment, facilitate access to financing, and create enabling environments for technology transfer and capacity building.

7.6.4 Global collaboration and knowledge sharing

Collaboration and knowledge sharing are essential for driving the future of renewable energy integration in agriculture. Governments, research institutions, agricultural organizations, and industry stakeholders will work together internationally to exchange best practices, insights, and technological advancements. Platforms that promote knowledge exchange, capacity building, and public–private partnerships will help spread information, expertise, and resources, creating a global community dedicated to promoting sustainable agriculture by integrating renewable energy. Ultimately, the future of incorporating renewable energy into agriculture is marked by a focus on innovation, collaboration, and sustainability.

REFERENCES

1. Mete Yavuz and Mer Cihan Kivanç, "Optimisation of a cluster-based energy management system using deep reinforcement learning without affecting prosumer comfort: V2X technologies and peer-to-peer energy trading," *IEEE Access*, vol. 12, pp. 31551–31575, 2024.
2. N. Singh, S. Tamrakar, A. Mewada, and S. K. Gupta, *Artificial Intelligence Techniques in Power Systems Operations and Analysis* (1st ed.), Auerbach Publications, pp. 1–72, 2023. https://doi.org/10.1201/9781003301820
3. N. Singh, A. K. Sharma, I. Sarkar, S. Prabhu, and K. Chadaga, "IoT-based greenhouse technologies for enhanced crop production: A comprehensive study of monitoring, control, and communication techniques," *Systems*

Science & Control Engineering, vol. 12, no. 1, 2024. https://doi.org/10.1080/21642583.2024.2306825

4. Ibrahim Hetita, Amr S. Zalhaf, Diaa-Eldin A. Mansour, Yang Han, Ping Yang, and Congling Wang, "Transient Overvoltage Response of Photovoltaic Panels to Lightning Impulse: Experimental Observations and Analysis," *2023 12th Asia-Pacific International Conference on Lightning (APL)*, pp. 1–6, 2023.
5. Nagendra Singh, Neeraj Priyadarshi, and Niraj Kumar, "Meta Heuristic Techniques Used to Improve the Performance of Static Converters," *International Conference on Innovative Development in Virtual and Rapid Prototyping (IDVRP-2022). AIP Conference Proceedings* 2771, 020027, 2023. https://doi.org/10.1063/5.0152913
6. Nagendra Singh and Yogendra Kumar, "Multi objective economic load dispatch problem solved by new PSO," *Advances in Electrical Engineering*, vol. 2015, pp. 2356–6655, 2015. https://doi.org/10.1155/2015/536040
7. Mohammad Jamjoum, Raneem Negry, Emad Abdelsalam, Fares Almomani, Tareq Salameh, and Ahmed Al Makky, "Customers Dependent Demand Response in Energy Hubs with IoT Perforation," *2023 Advances in Science and Engineering Technology International Conferences (ASET)*, pp. 1–5, 2023.
8. A. Nighoskar, P. Chaurasia, and N. Singh, "Advancing nuclear energy forecasting: Exploring regression modeling techniques for improved accuracy," *Nuclear Engineering and Technology*, vol. 57, no. 1, p. 103144, 2025. https://doi.org/10.1016/j.net.2024.08.013
9. Nagendra Singh, Tulika Chakrabarti, Prasun Chakrabarti, Vladimir Panchenko, Dmitry Budnikov, Igor Yudaev, and Vadim Bolshev, "Analysis of Heuristic Optimization Technique Solutions for Combined Heat–Power Economic Load Dispatch," *Applied Sciences*, vol. 13, no. 18, p. 10380, 2023. https://doi.org/10.3390/app131810380
10. Daniel Kris Hamonangan Sitompul, Oyas Wahyunggoro, and Adha Imam Cahyadi, "Single-Axis Solar Tracker for Solar Panel with Power and Cost Analysis," *2021 International Conference on Advanced Mechatronics, Intelligent Manufacture, and Industrial Automation (ICAMIMIA)*, pp. 18–22, 2021.
11. Nagendra Singh, Shekh Kulsum Almas, Ritesh Tirole, Amit Jain, Siddharth Shukla, and Yogendra Kumar, "Analysis of Optimum Cost and Size of the Hybrid Power Generation System Using Optimization Technique," *12th IEEE International Conference CSNT-2023 held at Technocrats Institute of Technology (Excellence)*, 8 & 9 April 2023.
12. Santy Saijeet, Sriparna Roy Ghatak, Sarat Chandra Swain, "Planning of Unbalanced Distribution Network Using Solar Energy Sources under Uncertain Environment," *2021 1st International Conference on Power Electronics and Energy (ICPEE)*, pp. 1–5, 2021.
13. Nagendra Singh, "Optimization techniques used for design of economic electrical power distribution," *Meta Heuristic Optimization Techniques*, Walter De Gruyter GmbH, pp. 79–92, 31 January2022. https://doi.org/10.1515/9783110716214-005

14. Vandana Sondhiya, Kaustubh Dwivedi, Shekh Kulsum Almas, and Nagendra Singh, "Energy management of nanogrid through flair of deep learning from IoT environments," *Artificial Intelligence Techniques in Power Systems Operations and Analysis*, Taylor& Francis, July 2023.
15. Nagendra Singh, Ritesh Tirole, Shekh Kulsum Almas, and Dimpy Sood, "Renewable power generation using asynchronous generator: A review," *Springer Lecture Notes in Electrical Engineering (Recent Advances in Power Systems)*, vol. 688, pp. 205–215, October 2020. https://doi.org/10.1007/978-981-15-7994-3_18

Chapter 8

A protection scheme based on SVM-kNN for pole-to-ground and pole-to-pole fault detection in DC microgrid

Shankarshan Prasad Tiwari

8.1 INTRODUCTION

In the current energy scenario, the demand for electrical energy is continuously increasing with a simultaneous increment in the electrical power appliance. Therefore, challenges toward continuous power supply to integrated load centers are rising. To generate electrical power to balance demand and generation, numerous bulk power generation sites have been installed in dissimilar geographical locations. However, fuel availability throughout the year and problems encountered in transportation restrict such power generation technology for future use in power sectors. The other problems with such power generation sites are the degradation of the atmospheric conditions due to the emission of dangerous air pollutants such as CO, CO_2, and SO_2. The adoption of microgrids is an alternate solution due to advancements in power semiconductor technology and its advantages over traditional power generation sites. The microgrid can be formed by amalgamation of the small-rated renewable energy resources (RESs), converters, load, and battery (energy management unit) with some pre-specified parameters or boundaries [1–3]. Based on the power output, three classes of microgrids are available: alternating current (AC), direct current (DC), and hybrid microgrids. In recent years, it has been observed that the application of smart DC-based appliances are used in dissimilar sectors, such as light-emitting diode (LED) lighting, electric vehicle (EV) charging, grinders, savers, and electronic gadgets based on DC loads. The adoption of DC-based power appliances has motivated the formation of the DC microgrid to provide DC power for efficient operation. There are a number of advantages of a DC microgrid over an AC-based power distribution system such as the capability of higher power transfer [4], effectiveness in cost [5], fewer power conversion stages, as well as a smaller number of connecting leads.

Nevertheless, the crucial task in the DC microgrid is protection against faulty conditions under varying faulty conditions. Rapid variations in the fault current can damage costly converter switches as well as other parts of

 DOI: 10.1201/9781003516156-8

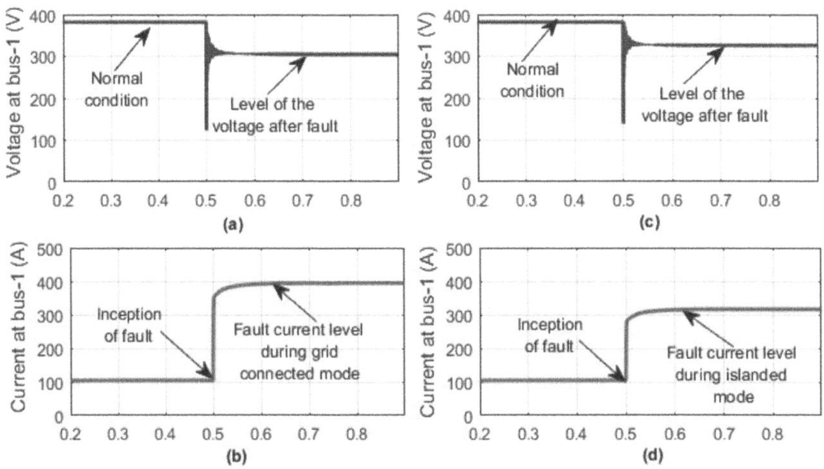

Figure 8.1 (a and b) Change in the voltage–current level during grid mode. (c and d) Variation in the voltage–current level islanded mode.

the system. To address the variation in the current level after the occurrence of the fault resistance, Figures 8.1 and 8.2 have been used, where it can be easily observed that the fault current level is changing in each case. Variations can be investigated in Figure 8.1(a) and (b), while in Figure 8.2, two cases of high fault resistance can be considered under both of the modes to show changes in current when fault resistance is changing. The protection approach should be efficient and capable of detecting such conditions rapidly to save the rest of the power distribution system or any type of distribution system. Researchers in the existing works have proposed a variety of protection schemes, some of which are considered here, such as the fuzzy granulation-based protection scheme for inverter-based microgrid in reference [6], where local bus voltage as well as positive pole current have been used for fault analysis in this work.

An adaptive scheme as well as an integrated fault detection and control scheme are described in references [7] and [8]. An entropy-based scheme in reference [9] uses Shannon entropy to evaluate the presence and absence of faulty conditions in the microgrid. Series arc fault detection in reference [10], where a Distributed Unknown Input Observers (DUIO) scheme has been used to identify arc faults in the microgrid. A novel work under voltage sensor faults is reported in reference [11], high-impedance fault detection schemes in references [12] and [13], where some dissimilar operating parameters have been considered when developing a scheme for fault identification. Using fewer sensors, a protection scheme for fault detection is

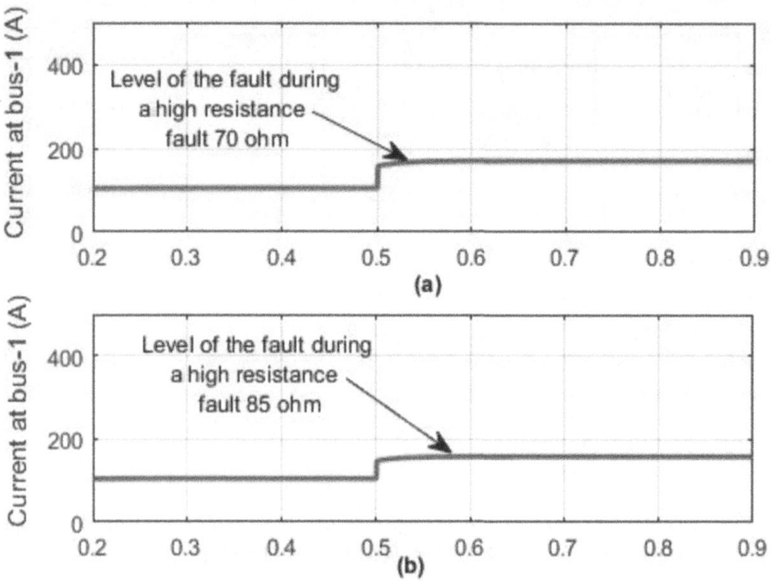

Figure 8.2 Variation in fault current level during high fault resistance.

described in reference [14], and to address communication failure, a protection scheme is described in reference [15], where differential protection with bus power gradient has been given. With the use of a voltage–current signal, a data-driven approach in reference [16], and a deep belief network (DBN)-based protection in reference [17], the concept of the variational mode decomposition (VMD), and the sea-horses optimizer algorithm have been used when analyzing faults in the microgrid. An O-Z-source breaker-based scheme is presented in reference [18], based on mathematical morphology, a protection scheme in reference [19], and based on inductance estimation, a protection scheme in reference [20].

However, protection schemes have not used a variety of scenarios when simulating fault conditions. Therefore, a support vector machine (SVM)-k-nearest neighbor (kNN)-based protection strategy has been developed in this chapter to protect the microgrid which is operating under such types of varying operating scenarios in the proposed DC microgrid.

The key contributions in the proposed work are given below:

1. Development of a protection scheme for fault identification under grid-connected and islanded mode.
2. Analysis under different categories of the operating conditions in the microgrid

3. Simulation of the proposed DC microgrid to show abrupt changes in the fault current levels.

The remaining part of the chapter is organized as follows: Section 8.2 provides a single-line diagram of the test DC microgrid model, and Section 8.3 addresses an overview of the given algorithm. Protection scheme development is covered in Section 8.4 and performance analysis in Section 8.5. Conclusion is given in Section 8.6.

8.2 SINGLE-LINE PRESENTATION OF THE DC MICROGRID

To clearly show the architecture of the 380V DC microgrid, Figure 8.3 has been used, where a complete microgrid can be easily observed with the location of the distributed energy resources (DERs). In the proposed microgrid system, two photovoltaic (PV) generators (125 kW each) based on renewable DERs are connected at bus B2 and B4, while a linear load is connected at bus B3. To integrate the utility grid from the rest of the microgrid, point

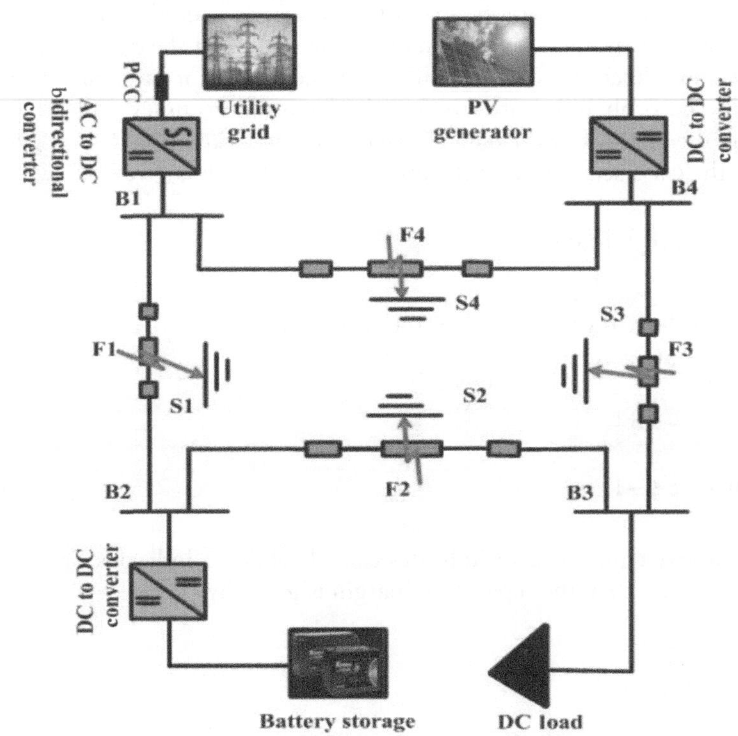

Figure 8.3 Single-line view of the test microgrid model.

of common coupling (PCC) is given. Therefore, the microgrid can be used either in grid-connected mode or isolated mode, where the grid is not connected to the microgrid. A constant impedance load is integrated at bus B3 with a capacity of 100 kW, and an energy storage device (0.4 kAh) at bus B2. In addition to the above, a total of four sections have been used to connect buses and DERs where the length of each section is 2 km. To analyze the behavior of the fault in the microgrid, a total of four faults have also been used with dissimilar fault resistances. An 11 kV/0.380 kV utility grid is connected to bus B1.

8.3 OVERVIEW OF THE ALGORITHM

8.3.1 Support vector machine

SVM is a well-known algorithm and has been widely accepted in recent years for the recognition of datasets. It works on the pattern recognition principle for the given class of data. It is basically a binary classifier that is best suited for two-class classification. In the proposed work, SVM has been used for the classification of the mode of operation using datasets. In SVM, for data classification, a set of hyperplanes is used [21,22], which is significant for the data separation task. The margin of the data points in the hyperplane determines the classification accuracy. For large-distance data points, the probability of error will be less, and accuracy will be greater during the analysis of the data classification task. Equation (8.1) is given to show the class separation of the given dataset, which is given below:

$$T = \left(a_i, b_i\right), < i < n. \tag{8.1}$$

Here a_i is the real vector and bi is the class of data. For two class data, below equations are given:

$$w.x - c \geq +1 \tag{8.2}$$

$$w.x - c \leq -1. \tag{8.3}$$

In the above equations, +1 indicates class 1 while −1 indicates class 2. If m is the margin, then the separation margin is given by:

$$m = \frac{2}{\|w\|} \tag{8.4}$$

m should be increased for a greater margin.

8.3.2 k-nearest neighbor

In recent years, the kNN has been widely accepted by researchers for analysis of the faulty conditions in power distribution systems. It works on the principle of the recognition of the available pattern of data. The algorithm finds kNNs in the feature space to imitate the algorithm for further operation. Based on the similarity index, distance is measured with the selection of the value of k that can be between the appropriate odd number [23]. To avoid tie situations during the prediction of the dataset, it is significant to select the odd value of the neighbors.

8.4 PROTECTION SCHEME AND OVERVIEW OF THE USED MODULES

A complete layout has been proposed in this section using SVM and kNN modules, where both operating modes have been used when developing the protection scheme layout. After a fault occurrence, rapid detection is necessary, which can be performed using efficient planning and quick action. The operation sequence of the protection scheme can be analyzed using Figure 8.4 for a clear operation of the protection scheme. First, samples of the electrical signals are recorded at selected buses and then used in the protection scheme for further tasks. After sample collection of the electrical signals, mode identification has been performed when the utility grid is in the system. Once the mode is identified, the algorithm initiates the remaining modules where the fault detection/classification task through the given modules in the protection scheme has been performed in grid as well as islanded mode. Section identification will be performed after a fault occurrence using dedicated modules. In the protection scheme, a total of five modules have been used, as given in Figure 8.4, where an SVM-based module is used for mode detection and kNN-based modules for reaming tasks, as illustrated in the below protection scheme.

For grid-connected mode, kNN-1 and kNN-2 are used, while in islanded mode, kNN-3 and kNN-4 are used in the protection scheme. Once the fault has been sensed and a faulty section is identified, a trip signal will be generated to initiate a relay operation to isolate the faulty and healthy parts in the microgrid. In healthy conditions, the algorithm is in its initial state and there is no operation signal for the operation of the breaker.

8.5 PERFORMANCE ANALYSIS

The robustness of the protection scheme can be increased by the accurate and efficient performance of the system. Therefore, identification is a necessary task to identify the accuracy of the microgrid. To generate data for training and testing, Table 8.1 has been used, which shows 2048 total fault

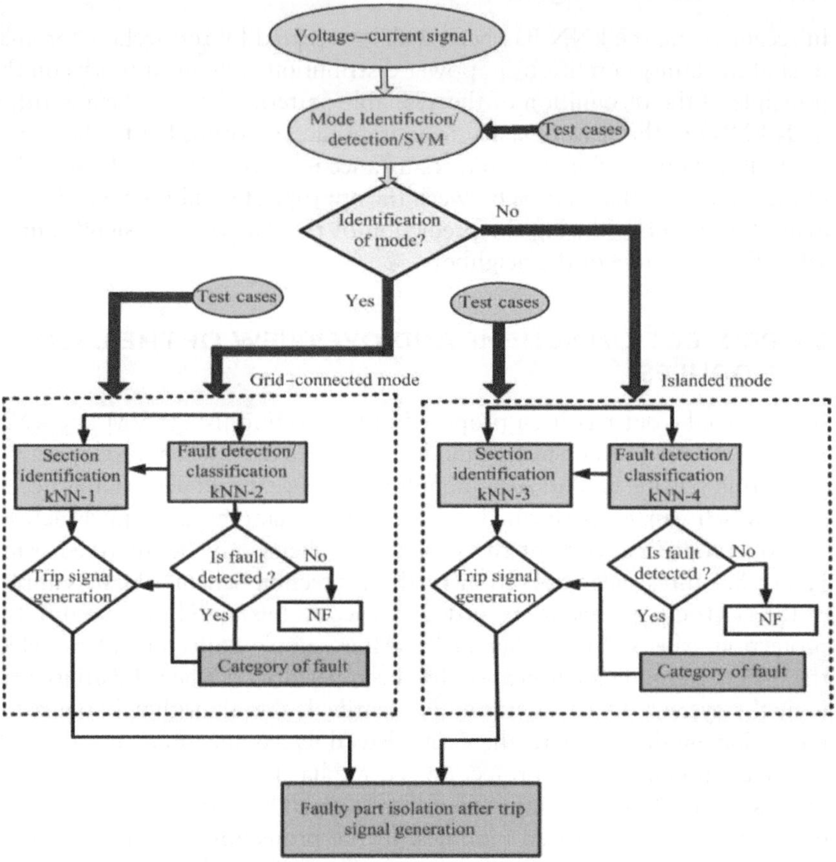

Figure 8.4 Protection scheme layout.

Table 8.1 Data-generation pattern

Used parameters in data generation	*Total cases used for data generation*	*Total-generated cases*	*Total cases used for training of the modules*	*Total cases used for testing of the modules*
Fault class	2	Total cases = 2 × 16 × 16 × 2 × 2 + 50 = 2048 + 50 = 2098	1468	630
Fault resistance (Ω)	16			
Length cases considered during fault (km)	16			
Operating mode	2			
Fault inception (0.5,0.6)	2			
No-fault cases	50			

cases through simulation of the microgrid model as demonstrated in Section 8.2. Apart from the above cases, 50 no-fault cases have also been generated through the simulation of the proposed DC microgrid system. For training and testing purposes, fault cases have been divided in the ratio of 70%–30%, where the former percentage is used during training and remaining during testing of the protection scheme. For a complete analysis of the protection scheme, subsections have been used in this section.

8.5.1 Mode detection module

As stated in Section 8.2, a microgrid can be operated in either mode (grid or islanded); therefore, a protection scheme should be capable of identifying the working mode of the DC microgrid. The main reason behind mode identification is the fault current level during both operating modes, which can be lower in the islanded or isolated mode of the microgrid. In this subsection, performance has been validated for the SVM-based mode detector module. The performance of the module for mode identification has been evaluated using 480 test cases. After successful testing of the module, responses are dealt with in Table 8.2, where it can be observed that the SVM-based mode detection module is efficiently performing the mode identification task. To observe the invulnerability of the proposed scheme with other techniques, performance has been compared with decision tree (DT) and linear discriminant analysis (LDA)-based algorithms where the same testing datasets have been used as used for SVM-based module. To show a demonstration of the achieved accuracy, Figure 8.5 is used, where variation in the achieved accuracy can be easily observed. The results in Table 8.2 show the accurateness of the proposed SVM module.

8.5.2 Fault detection/classification

After a fault occurrence, rapid fault detection is necessary to save the remaining parts of the microgrid. It helps in planning of future activities when analyzing fault-related activities. Therefore, the efficacy of the protection scheme should be greater for prompt identification. Performance has

Table 8.2 Performance of the SVM module

Types of algorithms	*Mode of operation*		*Accuracy (%)*
	Grid connected (%)	*Islanded mode (%)*	
SVM	99.18	98.27	98.72
DT	97.72	96.54	97.13
kNN	96.53	95.88	96.20

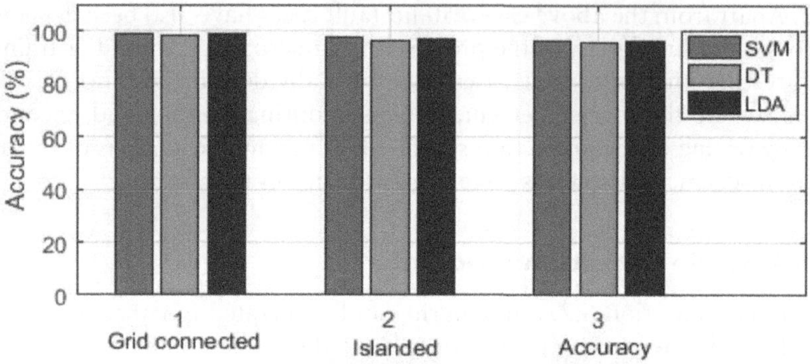

Figure 8.5 Comparison of the achieved accuracy of proposed mode detector module.

Table 8.3 Accuracy analysis of the fault detector with reliability indices

	Mode of operation					
	Grid connected			*Islanded*		
Algorithms	*Reliability indices-I (%)*	*Reliability indices-II (%)*	*Overall accuracy (%)*	*Reliability indices-I (%)*	*Reliability indices-II (%)*	*Overall (%)*
kNN	98.72	99.54	99.13	98.23	98.18	98.20
DT	96.52	97.94	97.23	95.49	96.37	95.39
LDA	95.88	96.29	96.08	94.28	95.88	95.08

been investigated through reliability analysis, where dependability as well as security analysis have been carried out in both operating modes. In existing literature, various studies are given where researchers have considered reliability analysis [24,25]. Dependability (reliability indices-I) is helpful in the analysis of the actual fault detection and the total given fault cases for detection. In security analysis (reliability indices-II), no-fault cases have been considered when analyzing security analysis. Security can be analyzed by detecting no-fault cases and giving no-fault cases. To validate the fault detection/classification module, a total of 565 test cases, including 15 no-fault cases, have been considered. After successful validations, results are dealt with in Table 8.3, where performance is higher for the kNN-based classifier. The achieved accuracy after successful testing is 99.13% and 98.20%, respectively, in both modes of the DC microgrid. The performance of the fault detector/classifier indicates that the protection scheme is effective when identifying faults in the DC microgrid. Figure 8.6 clearly demonstrates differences in achieved accuracy when compared with other algorithms.

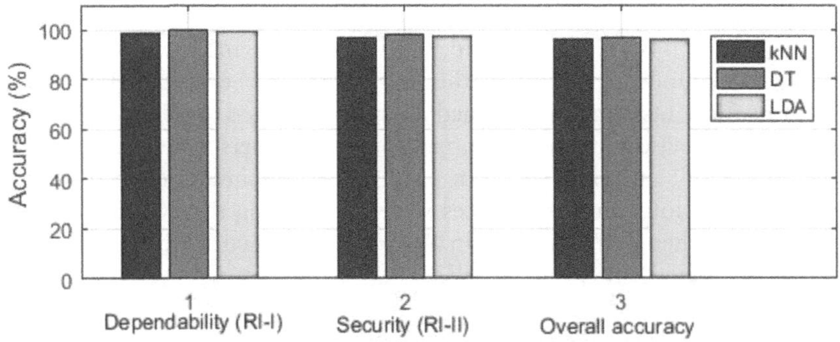

Figure 8.6 Performance of the fault detector under grid-connected mode.

Table 8.4 Performance during section identification using section identifier module

Mode	*Section/line*	*Type of fault*	*Fault resistance (Ω)*	*kNN (%)*
Grid connected	S-1	PG-1	75	98.74
	S-2	PG-2	80	98.92
	S-3	PP-1	0.4	98.36
	S-4	PP-2	0.5	99.12
	Overall accuracy			98.78
Islanded	S-1	PG-1	75	98.12
	S-2	PG-2	80	98.25
	S-3	PP-1	0.4	97.56
	S-4	PP-2	0.5	98.28
	Overall accuracy			98.05

8.5.3 Section identification

Faulty section or line identification after disturbance in the system due to the fault helps in the restoration of the power supply by removing the faulty part. Therefore, in this subsection, performance has been evaluated for section identification. To validate the overall analysis of the section identification module, testing has been performed through 240 test cases. The results in Table 8.4 indicate that performance is greater when identifying faulty sections through the section identification module.

8.6 CONCLUSION

The DC microgrid has several benefits over the traditional AC-based power distribution network, but protection is a major concern because of the huge changes in the fault current. This chapter presents the SVM-kNN algorithm to perform assigned tasks in the microgrid. Voltage and current samples were utilized in the development of the protection scheme in grid and islanded modes. In the proposed work, three main tasks have been performed: mode detection in the first stage; in the second stage, fault identification tasks are performed in both reported modes; and, in the third stage, section identification tasks have been performed for dissimilar sections. Fault current levels can vary due to dissimilar fault resistance; therefore, various fault resistances were used during the evaluation of the protection scheme. The results in Section 8.5 authenticate the efficacy of the protection scheme.

Acronyms

CO_2 carbon dioxide
DBN deep belief network
DER distributed energy resources
DT decision tree
DUIO Distributed Unknown Input Observers
EV electrical vehicle
kNN k-nearest neighbor
LDA linear discriminant analysis
LED light-emitting diode
PCC point of common coupling
PG pole to ground
PP pole to pole
PV photovoltaic
RESs renewable energy resources
SO_2 sulfur dioxide
SVM support vector machine
VMD variational mode decomposition

Symbols

% percentage
Ω ohm
km kilometer
kV kilovolt
kW kilowatt
S1–S4 sections in the microgrid

REFERENCES

1. Moslem Uddin, Huadong Mo, Daoyi Dong, Sondoss Elsawah, Jianguo Zhu, Josep M. Guerrero, "Microgrids: A review, outstanding issues and future trends," *Energy Strategy Reviews*, Vol. 49, pp. 101127, 2023.

2. Shankarshan Prasad Tiwari, "A dependable protection scheme for electrical vehicle integrated microgrid considering stressed fault scenarios and dissimilar fault inceptions," *Iranian Journal of Electrical & Electronic Engineering*, Vol. 20, no. 1, 2024.
3. Mohammad Bayat, Mohammad Ghiasabadi Farahani, Ali Asghar Ghadimi, Marcos Tostado-Véliz, Mohammad Reza Miveh, Francisco Jurado, "Optimal siting, sizing and setting of droop-controlled DERs in autonomous microgrids: A new paradigm in microgrid planning," *Electric Power Systems Research*, Vol. 225, pp. 109850, 2023.
4. Nagendra Singh, Sitendra Tamrakar, Arvind Mewada, Sanjeev Kumar Gupta, *Artificial Intelligence Techniques in Power Systems Operations and Analysis* (1st ed., pp. 1–72), 2023, Auerbach Publications. https://doi.org/10.1201/9781003301820
5. MdShafiul Alama, Fahad Saleh Al-Ismail, Syed Masiur Rahman, Md Shafiullah, Md Alamgir Hossain, "Planning and protection of DC microgrid: A critical review on recent developments," *Engineering Science and Technology, an International Journal*, Vol. 41, pp. 101404, 2023.
6. Hongyi Liu, Hua Han, Xinlong Zheng, Yao Sun Mei Su, Tao Ling, Xubin Liu, "Fuzzy granulation interval-based fault diagnosis method for ring-type DC microgrid," *IEEE Transactions on Smart Grid*, Vol. 15, no. 4, pp. 3402–3417, 2024.
7. Meysam Yadegar, Seyed Fariborz Zarei, Nader Meskin, Ahmed Massoud, "An adaptive voltage control compensator for converters in DC microgrids under fault conditions," *International Journal of Electrical Power & Energy Systems*, Vol. 156, pp. 109697, 2024.
8. Shekh Kulsum Almas, Nagendra Singh, "Power quality improvement for grid connected wind energy Generation system by implementation of DFIG scheme," *Turkish Online Journal of Qualitative Inquiry (TOJQI)*, Vol. 12, no. 9, pp. 6928–6944, 2021.
9. Reza Rahmani, Seyed Hossein Hesamedin Sadeghi, Hossein Askarian-Abyaneh, Mohammad Javad Emadi, "An entropy-based scheme for protection of DC microgrids," *Electric Power Systems Research*, Vol. 228, pp. 110010, 2024.
10. Shekh Kulsum Almas, Nagendra Singh, "Review on power quality improvement for grid connected wind energy generation system," *International Journal of Grid and Distributed Computing*, Vol. 13, no. 1s, pp. 106–112, July 2020.
11. Dat Thanh Tran, Myungbok Kim, Kyeong-Hwa Kim, "Seamless power management for decentralized DC microgrid under voltage sensor faults," *IEEE Access*, Vol. 11, pp. 74627–74640, 2023.
12. Meysam Yadegar, Seyed Fariborz Zarei, Nader Meskin, Frede Blaabjerg, "A distributed high-impedance fault detection and protection scheme in DC microgrids," *IEEE Transactions on Power Delivery*, Vol. 39, no. 1, pp. 141–154, 2024.
13. Xiaodong Wang, Ruojin Wang, Yingming Liu, Xing Gao, "Impedance ground faults detection and classification method for DC microgrid," *Journal of Electrical Engineering & Technology*, Vol. 18, no. 6, pp. 4011–4023, 2023.

14. Hua Han, Xinlong Zheng, Mei Su, Yao Sun, Zhangjie Liu, Hongyi Liu, Tao Ling, Xubin Liu, "A fast fault diagnosis scheme for ring bus DC microgrids with fewer sensors," *IEEE Transactions on Power Delivery*, 2023.
15. Weiliang Zhang, Hui Zhang, Na Zhi, "A novel protection strategy for DC microgrid considering communication failure," *Energy Reports*, Vol. 9, pp. 2035–2044, 2023.
16. Mohammad Daisy, Mahmood Hosseini Aliabadi, Shahram Javadi, Hasan Meyar Naimi, "Data-driven fault location approach in AC/DC microgrids based on fault voltage and current differences," *Sustainable Energy, Grids and Networks*, Vol. 36, pp. 101235, 2023.
17. Nagendra Singh, Neeraj Priyadarshi, Niraj Kumar, "Meta heuristic techniques used to improve the performance of static converters," In: *International Conference on Innovative Development in Virtual and Rapid Prototyping (IDVRP-2022), AIP Conference Proceedings* 2771, 020027, 2023. https://doi.org/10.1063/5.0152913
18. Zhong Zheng Zhou, Jiayan Jiang, Shu Ye, Dirui Yang, Jianguo Jiang, "Novel bidirectional O-Z-source circuit breaker for DC microgrid protection," *IEEE Transactions on Power Electronics*, Vol. 36, no. 2, pp. 1602–1613, 2021.
19. Navid Bayati, Hamid Reza Baghaee, Amin Hajizadeh, Mohsen Soltani, Zhengyu Lin. "Mathematical morphology-based local fault detection in DC Microgrid clusters," *Electric Power Systems Research*, Vol. 192, pp. 106981, 2021.
20. Morteza Shamsoddini, Behrooz Vahidi, Ramin Razani, Yasser Abdel-Rady I. Mohamed, "A novel protection scheme for low voltage DC microgrid using inductance estimation," *International Journal of Electrical Power & Energy Systems*, Vol. 120, pp. 105992, 2020.
21. Jenifer Mariam Johnson, Anamika Yadav, "Complete protection scheme for fault detection, classification and location estimation in HVDC transmission lines using support vector machines," *IET Science, Measurement & Technology*, Vol. 11, no. 3, pp. 279–287, 2017.
22. Ting Ke1, Xuechun Ge, Feifei Yin, Lidong Zhangc, Yaozong Zheng, Chuanlei Zhang, Jianrong Li, Bo Wang, Wei Wang, "A general maximal margin hyper-sphere SVM for multi-class classification," *Expert Systems with Applications*, Vol. 237, pp. 121647, 2024.
23. Shankarshan Prasad Tiwari, Ebha Koley, Subhojit Ghosh. "Protection scheme based on k-nearest neighbour algorithm for fault detection classification and section identification in DC microgrid," In: *Innovations in Electrical and Electronic Engineering: Proceedings of ICEEE*, Springer Singapore, Singapore, pp. 265–272, 2021.
24. Shankarshan Prasad Tiwari, "Fault detection in ring based smart LVDC microgrid using ensemble of decision tree," *Iranian Journal of Electrical & Electronic Engineering*, Vol. 18, no. 4, 2022.
25. Shankarshan Prasad Tiwari, Ebha Koley, Subhojit Ghosh, "Communication-less ensemble classifier-based protection scheme for DC microgrid with adaptiveness to network reconfiguration and weather intermittency," *Sustainable Energy, Grids and Networks*, Vol. 26, pp.100460, 2021.

Chapter 9

Calculation methodology and seasonality coefficients in determining power supply for rural consumers

Alina Vinogradova, Alexander Vinogradov, and Vadim Bolshev

9.1 INTRODUCTION

The functioning of technological equipment is carried out as well as labor and energy resources are used in accordance with the seasonality (by month and time of day) of the technological processes of many objects. As noted by various authors in works devoted to agricultural economics, management of agricultural production technologies, mechanization, and automation of technological processes, it is typical for agriculture, both for crop and livestock enterprises [1]. Agricultural facilities such as sugar factories, grain drying stations, vegetable storage facilities, greenhouses, and others have a pronounced seasonality of operation by month. In particular, for greenhouses, this applies to certain types of equipment, for example, lighting, heating, irrigation systems, and others [2]. The research [3,4] notes the influence of climatic factors (which have a pronounced seasonality) on the power consumption of various types of equipment, for example, oil production equipment. The seasonality of load schedules of rural electrical networks must be taken into account when forecasting power consumption. The study [4] shows that forecasting power consumption should take into account the seasonality of load schedules of electrical networks; the paper [5] does that the damage from power supply outages depends on the time of its onset and its duration.

Existing methodologies for calculating power supply reliability indicators (for example, Manov et al. [6] and Vinogradova et al. [7]) do not take into account the seasonality factor of power supply interruptions, considering the probabilities of failures in electrical networks to be constant throughout time. This does not allow more accurately predicting the power supply reliability for consumers, the likely damage from power supply interruptions, or/and developing measures to improve reliability [8, 9]. At the same time, a study of statistical data on outages in electrical networks of different voltage classes, carried out using the example of the Orel region, Russia, showed that the flow of failures and the causes of outages are seasonal [10].

DOI: 10.1201/9781003516156-9

9.2 DEVELOPED METHODOLOGY

Developed methodology allows calculating indicators of power supply reliability for consumers taking into account the seasonality of power supply outages by month and time of day. The methodology given in Vinogradov [8]was adopted as a prototype for the developed methodology. A network section refers to sections of power lines, as well as power sources and switching devices installed in the network.

The first step of the methodology is to determine statistical data on failures in the region's electrical networks.

Based on these data, the failure flow parameter, ω_{0i}, per 100 km of power lines or 100 equipment units, year^{-1} is determined by equation (9.1):

$$\omega_{0i} = \frac{N_{outages} \cdot 100}{n_{years} \cdot L_{\Sigma}}, \tag{9.1}$$

where $N_{outages}$ is the number of outages for the analyzed number of years, units;

n_{years} – number of years analyzed, year;

L_{Σ} is the total length of analyzed networks, km, or total number of analyzed equipment, pcs.

Also, from these data, the average power supply restoration time, $T_{restor.av.i}$, h, is determined using equation (9.2):

$$T_{restor.av.i} = \frac{\Sigma T_{outage.i}}{N_{outages}}, \tag{9.2}$$

where $\Sigma T_{outage.i}$ is the total restoration time for all power supply outages for the analyzed period of time, h.

If ω_{0i} and $T_{restor.av.}$ are not possible to determine from statistical data, both of them can be determined from reference or literary data, for example, Papkovand Osokin [5] and Manov et al. [6].

Next, it is necessary to find the value of seasonality coefficients characterizing the differences in failure flow indicators and restoration time for different months of the year and different hours of the day. The coefficients are determined separately for each voltage class of the networks under consideration.

The coefficient for taking into account the year seasonality by restoration time, $k_{T.year.j}$, dimensionless, is determined by equation (9.3):

$$k_{T.year.j} = \frac{12 \cdot t_{rest,\ month.j}}{\Sigma t_{rest,\ month.j}}, \tag{9.3}$$

where $t_{rest,\ month.j}$ is total restoration time after power supply outages in the *j*th analyzed month according to statistical data, h;

$\Sigma t_{rest,\ month.j}$ is total restoration time after power supply outages during the year according to statistical data, h; 12 is the number of months in a year.

The coefficient for taking into account the day seasonality based by restoration time, $k_{T.day.j}$, dimensionless, is determined by equation (9.4):

$$k_{T.day.j} = \frac{12 \cdot t_{rest,hour.j}}{\Sigma t_{rest,hour.j}}, \tag{9.4}$$

where $t_{rest,\ hour.j}$ is total restoration time after power supply outages that occurred in the analyzed hour in the *j*th hour according to statistical data, h;

$\Sigma t_{rest,\ hour.j}$ istotal restoration time after power supply outages during the day, according to statistical data, h; 24 is the number of hours in a day.

The coefficient for taking into account the year seasonality by the flow of failures, $k_{\omega.year.j}$, dimensionless, is determined by equation (9.5):

$$k_{\omega.year.j} = \frac{12 \cdot N_{outage.month.j}}{\Sigma N_{outage.month.j}}, \tag{9.5}$$

where $N_{outage.month.j}$ is the number of power supply outages in the *j*th analyzed month according to statistical data, pcs;

$\Sigma N_{outage.month.j}$ is the number of power supply outages during the year according to statistical data, pcs.

The coefficient for taking into account the day seasonality by the flow of failures, $k_{\omega.day.j}$, dimensionless, is determined by equation (9.6):

$$k_{\omega.day.j} = \frac{12 \cdot N_{outage.hour.j}}{\Sigma N_{outage.hour.j}}, \tag{9.6}$$

where $N_{outage.hour.j}$ is the number of power supply outages in the *j*th analyzed hour according to statistical data, pcs;

$\Sigma N_{outage.hour.j}$ is the total number of power supply outages during the day according to statistical data, pcs.

Next, it is necessary to determine the network length and configuration for which the calculation will be done. The next step is to determine the analyzed season. This can be a specific month, day, or a specified number of days in a given month, or several months, or a year.

The estimated number of hours of emergency power supply outages on each network section for a given season, $T_{outage.emer.sec.i.S}$, h, is determined by equation (9.7):

$$T_{outage.emer.sec.i.S} = \frac{T_{restor.av.i} \bullet \omega_{0i} \bullet l_i}{100} \bullet \sum_{1}^{n_{ma}} \left(\frac{k_{T.year.j} \cdot k_{\omega.year.j} \cdot n_{dma.j}}{n_{dm.j}} \right), \quad (9.7)$$

where l_i is the length of section *i*, km, or the amount of equipment, pcs;

n_{ma} is the number of analyzed months, units;

$n_{dma.j}$ is the number of analyzed days in the analyzed specific month, units;

$n_{dm.j}$ is the number of days in the analyzed specific month, units.

The number of hours of interruption from emergency power supply outages on a network section that occurred during a specific hour, $T_{outage.emer.sec.i.Sh}$, h, is determined by equation (9.8):

$$T_{outage.emer.sec.i.Sh} = \frac{T_{restor.av.i} \bullet k_{T.year.j} \bullet k_{\omega.year.j} \bullet \omega_{0i} \cdot l_i \cdot k_{T.day.j} \cdot k_{\omega.day.j}}{100}, \quad (9.8)$$

where $k_{T.year.j}$ is coefficient for taking into account the year seasonality by restoration timefor the analyzed month and voltage classes, dimensionless;

$k_{\omega.year.j}$ is coefficient for taking into account the year seasonality by the flow of failures for the analyzed month and voltage classes, dimensionless;

$k_{T.day.j}$ is coefficient for taking into account the day seasonality based by restoration timefor the analyzed specific hour of the day and voltage classes, dimensionless;

$k_{\omega.day.j}$ is coefficient for taking into account the day seasonality by the flow of failuresfor the analyzed specific hour of the day and voltage class, dimensionless.

The next step of the methodology is to determine the probability of failure-free operation (PFFO) of network sections during a given season, $P_{(PFFO.sec.i.S)}$, dimensionless. It is determined by equation (9.9):

$$P_{(PFFO.sec.i.S)} = \frac{t_{h.S} - T_{outage.emer.sec.i.S}}{t_{h.S}}, \tag{9.9}$$

where $t_{h.S}$ is the number of hours in the season, hours.

The PFFO of the entire network involved in the power supply of a given area is calculated in a given season taking into account its configuration. To do this, a network replacement diagram is constructed, which indicates the PFFO of each section for a given season. The network diagram is simplified by reducing series and parallel connected sections of the network to one element. PFFO of serial connected sections for a given season, $T_{(t).ser.i.S}$, is determined by equation (9.10):

$$P_{(t).ser.i.S} = \prod_{i=1}^{n_{ser}} {}_{(PFFO.sec.i.S)}(t_C) = e^{-t\sum_{i=1}^{n_{ser}}\lambda_i}, \tag{9.10}$$

where e is the base of the natural logarithm;

t is time under consideration, year;

λ_i is failure rate per 100 km of line, or per 100 pcs equipment (it is equal to the failure rate ω_{0i} in the case under consideration), year^{-1};

n_{ser} is number of series-connected network sections in the circuit under consideration;

$P_{(PFFO.ser.i.S)}$ is the PFFO for the *i*th network section taking into account the season.

PFFO of parallel connected sections for a given season, $T_{(t).par.i.S}$ (for the number of parallel sections n_{par}), is determined by equation (9.11):

$$P_{(t).par.i.S} = 1 - \prod_{i=0}^{n_{par}} (1 - P_{(PFFO.sec.i.S)}(t_C)), \tag{9.11}$$

Gradually, the equivalent circuit is reduced to 1 element which is the PFFO of the entire network.

Subsequently, indicators of power supply outages associated with maintenance and scheduled repairs of network sections are determined. As a rule, such an assessment takes into consideration a given value for the number of planned repairs for each type of network equipment. However, when studying regional power systems and their constituent electrical networks, it is not always possible to assess reliability indicators associated with planned outages by this method. Also, the number of planned repairs does not always coincide with the actual one. In addition, some power grid companies are switching to servicing equipment "on a condition-by-condition basis," meaning that some equipment will no longer receive scheduled repairs in quantities that match the standard value. Therefore, it is rational to rely on statistical data on the number and duration of planned outages.

For this purpose, statistical data on planned power supply outages in the regions electrical networks is collected. Based on them, the flow parameter of planned outages, m_{0i}, per 100 km of power lines, or per 100 pcs equipment, year^{-1} is determined according to equation (9.12):

$$\mathrm{m}_{0i} = \frac{N_{pl.outage} \cdot 100}{n_{a.year} \cdot L_{\Sigma}}, \tag{9.12}$$

where $N_{pl.outage}$ is the number of planned outages for the analyzed number of years, units;

$n_{a.year}$ is the number of analyzed years, year;

L_{Σ} is the total length of analyzed networks, km, or total number of analyzed equipment, pcs.

Also, from these data, the average time of planned outages, $T_{pl.outage.av.i}$, h, is determined by equation (9.13):

$$T_{pl.outage.av.i} = \frac{\Sigma T_{pl.outage.i}}{N_{pl.outage}}, \tag{9.13}$$

where $\Sigma T_{pl.outage.i}$ is the total time of planned outages for the analyzed period of time, hours.

If m_{0i}, and $T_{pl.outage.av.i}$ are not possible to determine statistically, then both of them can be determined from reference data, for example from Papkov and Osokin [5] (taking into account that this source provides values m_{0i} and $T_{pl.outage.av.i}$ for 1unit of equipment or for one line). In this case, m_{0i}, and $T_{pl.outage.av.i}$ are denoted as $m_{0i.ref}$, and $T_{pl.outage.av.i.ref}$ and the time of planned power supply outages in the network section, $T_{pl.outage.sec.i}$, h/year will be determined by equation (9.14):

$$T_{pl.outage.sec.i} = T_{pl.outage.av.i.ref} \cdot m_{0i.ref} \tag{9.14}$$

The next step is the determination of the value of seasonality coefficients characterizing the differences in the flow of planned outages and the time of planned outages for different months of the year and different hours of the day.

Coefficients taking into account the seasonality of the year based on the time of planned outages, $k_{T.year.av.j}$, $k_{m.year.j}$, $k_{T.day.av.j}$, $k_{m.day.j}$ are determined similarly to the seasonality coefficients for emergency outages, taking into account the fact that when calculating them, data on the number and time of planned outages is used.

The estimated number of hours of planned outages at each network section for a given season, $T_{pl.outage.sec.i.S}$, h/sec, is determined by equation (9.15):

$$T_{pl.outage.sec.i.S} = \frac{T_{pl.outage.av.i} \cdot m_{0i} \cdot l_i}{100} \cdot \sum_{1}^{n_{ma}} \left(\frac{k_{T.year.av.j} \cdot k_{m.year.j} \cdot n_{dma.j}}{n_{dm.j}} \right), \tag{9.15}$$

where l_i is the length of section *i*, km, or the amount of equipment, pcs;

n_{ma} is the number of analyzed months, units;

$n_{dma.j}$ is the number of analyzed days in the analyzed specific month, units;

$n_{dm.j}$ is the number of days in the analyzed specific month, units.

In the case of using reference data $m_{0i.ref}$, and $T_{pl.outage.av.i.ref}$, the value of $T_{pl.outage.sec.i.S}$ is determined by equation (9.16):

$$T_{pl.outage.sec.i.S} = T_{pl.outage.av.i.ref} \cdot m_{0i.ref} \cdot l_i \cdot \sum_{1}^{n_{ma}} \left(\frac{k_{T.year.av.j} \cdot k_{m.year.j} \cdot n_{dma.j}}{n_{dm.j}} \right) \quad (9.16)$$

The number of hours of interruption from planned outages on a network section that occurred during a specific hour, $T_{pl.outage.sec.i.Sh}$, h, is determined by equation (9.17):

$$T_{pl.outage.sec.i.Sh} = \frac{T_{pl.outage.av.i} \cdot k_{T.year.av.j}, \cdot k_{m.year.j} \cdot m_{0i} \cdot l_i \cdot k_{T.day.av.j} \cdot k_{m.day.j}}{100} \quad (9.17)$$

In the case of using reference data $m_{0i.ref}$, and $T_{pl.outage.av.i.ref}$, the value of $T_{pl.outage.sec.i.Sh}$ is determined by equation (9.18):

$$T_{pl.outage.sec.i.Sh} = T_{pl.outage.av.i.ref} \cdot k_{T.year.av.j}, \cdot k_{m.year.j} \cdot m_{0i.ref} \cdot l_i \cdot k_{T.day.av.j} \cdot k_{m.day.j} \quad (9.18)$$

The next step of the methodology is to determine the PFFO of network sections based on planned outages during a given season, $P_{(PFFO.pl.sec.i.S)}$, dimensionless. It is determined by equation (9.19):

$$P_{(PFFO.pl.sec.i.S)} = \frac{t_{h.S} - T_{pl.outage.sec.i.Sh}}{t_{h.S}} \quad (9.19)$$

Next, the PFFO of the entire studied network involved in the power supply of a given area is calculated, taking into account its configuration in a given season for both emergency and planned outages. To do this, a network replacement diagram is constructed, which indicates the PFFO for both emergency and planned (PFFOPL)outages of each section for a given season.

The calculation is carried out taking into account the following situation: during repair or maintenance of one of the consumer power circuits, the second (backup) is not included in the scheduled repair.

Therefore, when calculating redundant circuits, the PFFOPL of one circuit and the PFFO of the second one are taken into account; e.g., the probability of an emergency shutdown occurring in the reserve circuit is determined during the planned repair of the first.

The network diagram is simplified by bringing series- and parallel-connected sections of the network to one element for PFFOPL of series-connected sections for a given season, $P_{(t).ser.pl.S}$, by equation (9.20):

$$T_{(t).ser.pl.S} = \prod_{i=1}^{n_{ser.pl}} {}_{(PFFO.pl.sec.i.S)}\left(t_C\right) = e^{-t \sum_{i=1}^{n_{ser.pl}} \cdot m_{0i}}, \tag{9.20}$$

where e is the base of the natural logarithm;

t is time under consideration, year;

$n_{ser.pl}$ is number of series-connected network sections;

$P_{(PFFO.pl.sec.i.S)}$ is PFFOPL for the ith network section taking into account the season.

For two parallel sections (reserve) of the PFFO according to planned and emergency outages, $P(t)_{nap.n\pi.a\beta C}$, is determined by equation (9.21):

$$T_{(t).par.pl.emer.S} = P_{(PFFO.pl.sec.i.S)} + P_{(PFFO.sec.i.S)} - P_{(PFFO.pl.sec.i.S)} \times P_{(PFFO.sec.i.S)} \tag{9.21}$$

Gradually, the equivalent circuit is reduced to 1 element which is the PFFO of the entire network (for planned and emergency outages).

9.3 CASE STUDY

The case study for the developed methodology is based on statistical data on failures in rural electrical networks of 10–110 kV [7]. During the case study, the seasonality coefficients for failures were calculated taking into account the causes of outages on the specified networks (for the conditions of the Oryol region) by month and time of day. The calculation results are given in Tables 9.1–9.4.

On 110 and 35 kV networks, outages were recorded from April to September; they did not occur at other times of the year. Their causes are, as a rule, climatic factors such as thunderstorms, wind, etc. This is reflected by the seasonality coefficients for both the flow of failures and power supply restoration time for a given class of networks. The values of seasonality coefficients for 0.4–10 kV networks show less dependence on the seasons

Table 9.1 Seasonality coefficients for failure flow in 0.4–110 kV electrical networks (for the conditions of the Oryol region) by month of year

	Month											
	January	*February*	*March*	*April*	*May*	*June*	*July*	*August*	*September*	*October*	*November*	*December*
Voltage class, kV	*1*	*2*	*3*	*4*	*5*	*6*	*7*	*8*	*9*	*10*	*11*	*12*
110 kV (2018–2021)	–	–	–	1.6	2.4	0.8	2.4	3.2	0.8	–	0.8	–
35 kV (2019–2021)	–	–	–	–	–	3.4	3.4	1.7	3.4	–	–	–
10 (10.5) kV (2018–2021)	0.7	1	0.7	0.8	2.3	1.3	1.2	1.4	0.8	0.6	0.6	0.6
0.4 kV (2018–2021)	0.7	0.8	0.8	0.9	1.4	1.4	1.6	1.2	1	0.5	0.7	0.9

Table 9.2 Seasonality coefficients for power supply restoration time in 0.4–110 kV electrical networks (for the conditions of the Oryol region) by month of year

	Month											
	January	*February*	*March*	*April*	*May*	*June*	*July*	*August*	*September*	*October*	*November*	*December*
Voltage class, kV	*1*	*2*	*3*	*4*	*5*	*6*	*7*	*8*	*9*	*10*	*11*	*12*
110 kV (2018–2021)	–	–	–	0.5	3.3	0.3	2.1	4	1.6	–	0.2	–
35 kV (2019–2021)	–	–	–	–	–	4.49	3.02	3.3	1.19	–	–	–
10 (10.5) kV (2018–2021)	0.63	0.81	1.12	1.26	1.59	1.38	1.59	1.8	0.75	0.51	0.28	0.28
0.4 kV (2018–2021)	0.7	1	0.9	0.8	1.5	1.6	1.6	1.1	1.1	0.5	0.5	0.6

Table 9.3 Seasonality coefficients for failure flow in 0.4–110 kV electrical networks (for the conditions of the Oryol region) by time of day

	Hour											
Voltage class, kV	*0*	*1*	*2*	*3*	*4*	*5*	*6*	*7*	*8*	*9*	*10*	*11*
110 kV (2019–2021)	–	–	–	–	1.6	1.6	–	–	1.6	1.6	1.6	–
35 kV (2019–2021)	–	–	–	–	3.4	–	–	–	–	–	–	–
10 (10.5) kV (2019–2021)	0.75	0.25	0.5	0.69	0.69	1	1.1	0.97	1.7	1.2	0.85	0.97
0.4 kV (2018–2021)	0.35	0.25	0.14	0.1	0.2	0.14	0.25	0.5	1.1	1.3	1.5	1.15

	Hour											
Voltage class, kV	*12*	*13*	*14*	*15*	*16*	*17*	*18*	*19*	*20*	*21*	*22*	*23*
110 kV (2019–2021)	–	–	1.6	–	3.2	4.8	1.6	1.6	–	1.6	1.6	–
35 kV (2019–2021)	3.4	–	6.9	–	3.4	3.4	–	–	–	3.4	–	–
10 (10.5) kV (2019–2021)	1.03	1.35	1.29	1.22	1.47	0.72	1.19	0.85	1.03	1.69	0.69	0.75
0.4 kV (2018–2021)	1.2	1.55	1.6	1.6	2.1	1.8	1.7	1.4	1.25	1.4	0.7	0.5

Table 9.4 Seasonality coefficients for power supply restoration time in 0.4–110 kV electrical networks by time of day

	Hour											
Voltage class, kV	*0*	*1*	*2*	*3*	*4*	*5*	*6*	*7*	*8*	*9*	*10*	*11*
110 kV (2018–2022)	–	–	–	–	0.7	3	–	–	–	3.7	0.5	–
35 kV (2019–2021)	–	–	–	–	5.31	–	–	–	–	–	–	–
10 (10.5) kV (2019–2021)	0.51	0.13	0.36	0.44	0.78	2.25	0.64	0.65	1.78	0.6	1.08	1.24
0.4 kV (2018–2021)	0.2	0.3	0.09	0.06	0.2	0.1	0.3	0.7	1.5	0.4	1.6	1.3

	Hour											
Voltage class, kV	*12*	*13*	*14*	*15*	*16*	*17*	*18*	*19*	*20*	*21*	*22*	*23*
110 kV (2018–2022)	–	–	–	–	4.3	6	0.6	4.2	–	0.6	0.4	–
35 kV (2019–2021)	4.4	–	7.88		1.65	1.28	–	–	–	3.48	–	–
10 (10.5) kV (2019–2021)	1.38	1	0.84	1.61	1.47	1.03	1.36	1.16	1.02	1.57	0.72	0.38
0.4 kV (2018–2021)	1.1	1.6	1.7	1.6	1.8	1.8	1.7	1.5	0.9	1.1	0.5	0.4

of the year, but, nevertheless, they are also of greater importance in the summer months.

Based on the time of day, it can be noted that the maximum flow of failures for 0.4 kV networks occurs in the period from 9 to 21 hours, with the highest seasonality coefficient for both the flow of failures and power supply restoration time around 4 p.m. The maximum flow of failures for 10 kV networks occurs from 7 a.m. to 6 p.m., with a peak at 8 a.m. The highest values of the seasonality coefficients (for the flow of failures up to 4.5, for restoration time up to 5.3) are observed for 110 kV networks in the summer months in the morning hours (4–5 a.m.). For 35 kV networks, the maximum seasonality coefficients are observed at 4 a.m. and 2 p.m..

9.4 CONCLUSION

Emergency and planned outages in electrical networks are seasonal by month of the year and time of day. This is due, in the case of emergency shutdowns, to the seasonality of climatic conditions, uneven load patterns, and other factors. In the case of planned outages, seasonality is caused by work schedules during the day, planning of repair programs, etc.

Existing methods for calculating power supply reliability indicators do not take into account the seasonality factor of power supply outages, considering the probabilities of outages occurring in electrical networks to be the same at different points in time. It leads to the impossibility of differentiated determination of the PFFO and the time of interruptions in power supply during given seasons. In turn, many consumers have a seasonal nature of the technological process and the damage from interruptions in power supply varies depending on the moment of the outage including planned ones.

The developed methodology makes it possible to calculate indicators of the reliability of power supply to consumers, taking into account the seasonality of interruptions in power supply by month and time of day. It involves determining the appropriate seasonality coefficients and adjusting the average annual reliability indicators of the network under consideration.

REFERENCES

1. Ukhanova V. Yu. "Automated control of the microclimate of greenhouses," *Electrical Technology and Equipment in the Agro-Industrial Complex*, Vol 68, no 1, pp. 68–72, 2021.
2. Gilev D.V., Eltyshev D.K. "Regression analysis of the influence of climatic factors on power consumption of oil production facilities," *Bulletin of the Perm National Research Polytechnic University. Electrical Engineering, Information Technologies, Control Systems*, Vol 35, pp. 152–168, 2020.
3. Stashkevich E.V., Ilyukhin I.G. "Forecasting and managing power consumption of microgrid communities using artificial intelligence," *Bulletin of the South Ural State University. Series: Energy*, Vol 22, no. 2, pp. 18–29, 2022.
4. Kulikov A.L., Papkov B.V., Sharygin M.V. "Analysis and assessment of the consequences of disconnecting electricity consumers," *Library of Electrical Engineering*, Vol 8, pp. 1–84, 2014.
5. Papkov B.V., Osokin V.L. *Probabilistic and Statistical Methods for Assessing the Reliability of Elements and Systems of the Electric Power Industry: Theory, Examples, Tasks*, Stary Oskol, Russia, 2022. TNT, 426p.
6. Manov N.A., Khokhlov M.V., Chukreev Yu. Ya., Shumilova G.P., Uspensky M.I., Chukreev M. Yu., Polubotko D.V., Gotman N.E., Startseva T.B. *Methods and Models Research into the Reliability of Electrical Power Systems*, Syktyvkar, Russia: Komi Scientific Center of the Ural Branch of the Russian Academy of Science, 2010, 292 p.
7. Vinogradova A.V., Lansberg A.A., Vinogradov A.V. *Energy System of the Oryol Region: Review of Statistical Information*, Orel, Russia, 2023 Kartush LLC, 360 p.
8. Vinogradov A.V. *Principles of Configuration Management of Rural Electrical Networks and Technical Means of Their Implementation*, Orel, Russia, 2023. Kartush LLC.
9. Singh N., Tamrakar S., Mewada A., Gupta S.K. *Artificial Intelligence Techniques in Power Systems Operations and Analysis* (1st ed.), 2023. New York, USA: Auerbach Publications, pp. 1–72.
10. Singh N., Mishra A., Gupta M., Dwivedi K., Swarnkar M. Implementation of Microgrid for Rural Electrification to Cater on Side Load Demand, Wuhan, China, *2023 3rd International Conference on Energy, Power and Electrical Engineering (EPEE)*, pp. 1058–1061.

Chapter 10

Application of advanced sensors for monitoring in power system device

Dipti Shukla

10.1 INTRODUCTION

Generating, distributing, and transmitting systems are all part of the power system network. It transforms energy into electrical energy by using the energy's form (such as coal or diesel).The power is sent to the several substations via the transmission line. Power is sent from the substation to the distribution transformer, which steps it down to a level that is adequate for the customers [1]. Distributing and transmitting electricity is a crucial component of the electric power system. Delivering electricity to the final consumer is its primary goal.

The majority of them are situated far away from the end-user on a radial topology with generating. The transmission and distribution system mostly falls short in offering service providers, operators, and end users real-time data monitoring. Utilities can regulate how much electricity is used because of advanced sensor technology [2]. It is a clever method by which the provider may improve overall stability and efficiency as well as make the transmission and distribution of electric power more suitable. In recent times, scientists have been striving to increase the intelligence of the electrical power system through the utilization of sophisticated sensors for the three processes of power transmission and distribution monitoring [3]. Research and development on sensors is a daily activity that produces a great deal of intriguing and significant goods [4].

New materials and electrical systems that are small, smart, and efficient, as well as micro- and nanotechnology, will all have a significant impact on future sensors. Sensors affect every aspect of human life and can also be used to improve the world in ways such as improved medical diagnostics, improved environmental monitoring, improved performance of energy sources such as solar power, fuel cells, and batteries, improved power systems, improved human health and safety, and improved space exploration. These days, potentiometers and force-detecting resistors are typical conventional sensors. However, because of developments in micromachining

DOI: 10.1201/9781003516156-10

and technology, producers can now make a lot more sensors at much quicker rates and with higher sensitivity [5].

Energy efficiency may be increased while costs are decreased by using novel techniques for the usage of sensors, sensor systems, and sensor data processing. As energy is necessary for contemporary business but is sometimes seen as an added expenditure, cost and performance are typically crucial considerations when applying sensor technology to industry [6]. Consequently, in most situations, cost-effectiveness is essential. Therefore, using sensor technologies in the real world requires adopting the right technology to overcome obstacles in this sector, such as the techno-economic analysis for the suggested system, creative approaches for the usage of current sensors, data analysis utilizing artificial intelligence (AI), big-data analysis, and operation optimizations [7].

In the Internet of Things (IoT) sensor, the advanced sensor is also a fundamental component. The IoT allows nearly anything to have a unique identity and to send data to and from those items via an analogous sensor network or the Internet. A wireless sensor and actuator network, which may comprise thousands of nodes, is one use for smart sensors. All the nodes are linked to individual actuators and one or more additional sensors and sensor hubs [8].

10.2 ELECTRIC POWER MONITORING AND DISTRIBUTION BY IOT SENSOR

IoT sensors are essential to the IoT, or the network of smart gadgets. These sensors gather critical data that can be transmitted in real time to consumers by being connected to gadgets such as thermostats, doorbells, vehicles, refrigerators, and smartphones. Devices with IoT sensors are being used by cities, municipalities, and buildings more often to track, evaluate, and control traffic, weather, heating, and cooling systems, as well as resources [9]. A crucial element of the electric power system is the distribution and transmission of electricity. Delivering electricity to the final customer is its primary goal. The majority of them have radial architecture and are situated far away from the end-user at a distant location [10].

For the most part, the transmission and distribution system falls short of offering service providers, operators, and end-users real-time data monitoring. Utilities can regulate how much electricity is used because of smart grid systems [11]. The provider may improve overall efficiency and stability as well as the transmission and distribution of electric power in a more suitable method through this clever technique. Researchers have been working to advance the intelligence of the electrical power system through power transfer and applying the IoT in smart grid distribution monitoring.

A smart grid can be defined as essentially integrating advanced control methods, intelligent devices, and information technology (IT) with the power system networks that already exist [12]. This allows power to be generated, transmitted, and distributed in a way that is more responsive, attractive, robust, and communicative. Reduced maintenance and operating costs, greater large-scale renewable energy utilization, decreased outages and disturbances, minimal power loss, enabled energy management systems, and real-time power flow monitoring are the key benefits of implementing a smart grid for utilities. In line with this, customer energy management systems and enabling electricity usage monitoring are two benefits of smart grid deployment for consumers [13].

New methods of data collection and information processing, referred to as IoT sensing technology, are required to combine these systems. The answer to the smart grid's need for seamless connectivity and integration is the IoT. Very few in-depth studies on power transmission and distribution monitoring systems utilizing IoT are currently conducted in Indonesia; the majority of studies primarily focus on the broad theoretical notion of a smart grid [14]. The technological aspects of IoT sensors for monitoring distribution and transmission that Indonesia may use are the focus of this article.

10.2.1 Smart grid architecture utilizing the IoT

Figure 10.1 shows a comprehensive smart grid system requires a variety of IoT components. IoT has drawn a lot of attention from academics recently because of its ability to analyze and exchange data in a variety of areas related to our daily lives [15]. The application layer, network layer, cloud management layer, and device layer make up the four levels of the IoT system architecture, as shown in Figure 10.1.

Data collection, processing, and computation from end-users are handled by the "application layer," which is the top layer. Applications like demand response management, visual monitoring platforms, real-time energy management, and other fundamental services are provided to customers by the application layer's vast array of intelligent application infrastructures [16]. Using the user interface, the application layer facilitates end-user interaction with the smart grid and reacts to data furnished by relevant devices and software descriptions. "Cloud management layer" is the name of the second layer.

Digital data may be remotely managed, processed, backed up, and maintained via cloud storage, which is part of the cloud service layer. Anytime the customer uses internet services, they may access and retrieve data from this cloud tier [17]. Users may make, modify, and oversee their account and authentication using its help. Data processing, computation, data management, and storage are the major components of the cloud management layer. "Network layer" is the third layer. In a smart grid, the network layer connects various devices to the application and device layers.

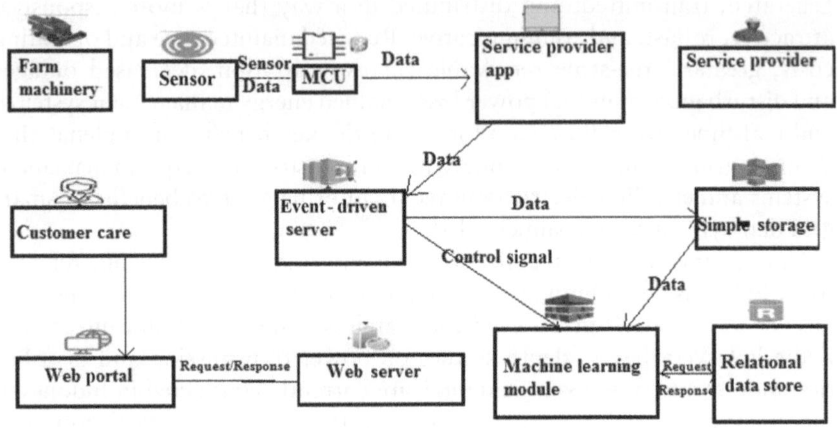

Figure 10.1 The architecture system of IoT.

Communication networks include the internet, fiber optic cable, power line communication, 3G/4G mobile communication networks, and wireless broadband, which make up the network layer. The objects sub-layer and gateway sub-layer are located in the "device layer," which is the fourth tier [18].

The objects layer is in charge of gathering raw data from actuators, smart plug-in devices, smart meters, and sensing apparatus. While this is going on, a microcontroller, communication module, and local storage are used by the gateway sub-layer to link all of the things layer devices and mediate numerous access points. The primary responsibilities of the device layer are to use a range of smart devices to collect, identify, process, and analyze all relevant data before sending it to the cloud management layer.

10.2.2 Important IoT technology for smart grids

A transmission and distribution monitoring system has to include several critical technologies to deploy IoT for smart grid applications. It may be explained as follows.

10.2.2.1 Smart sensor technology (SST)

Among the IoT's most crucial components for the smart grid is the smart sensor. An intelligent sensor is a gadget that provides information to the control system regarding definite characteristics and real events in the monitored physical item. Raw data is provided by smart sensor for information processing, feedback, and analysis. Several cutting-edge sensor technologies are now being used in several sectors [19].

Smart sensors in transmission and distribution monitoring systems can be employed in a variety of locations and dispersed over the grids, including substations, distribution transformers, transmission towers, and transmission and distribution cables. Two kinds of smart sensors are available for use. Initially, they are employed to measure the meteorological conditions surrounding the assets, including rain, humidity, wind speed, temperature, and direction sensors [5].

The network's ability to send and transfer data is aided by the second level's General Packet Radio Service (GPRS) module. Additionally, more reliable and efficient than previous technologies, GPRS can enhance network performance for real-time data. The GPRS module may also transfer data directly to the server by utilizing a specific node's address and the TCP or IP protocol. Data collection and transmission to servers or cloud management layers are made more dependable by employing two-layer wireless sensor networks [20].

10.2.2.2 Hardware design for transmissions

The IoT-based equipment deployment in the gearbox system is shown in Figure 10.2. It is made up of four components: the workstation, server (cloud management), data transmitter, and wireless sensor network. The transmission line and towers' physical environment may be sensed by wireless sensor network (WSN). Sensing data is gathered by sensors and sent between nodes or between nodes and servers via an IEEE 802.15.4 / Zigbee data transmitter module. After that, the combined data is sent to GPRS to provide WSN and the server with long-distance, high-speed information transfer.

All sensor data is collected, processed, analyzed, and stored by the server via a GPRS connection module. To receive and send data to the workstation, cloud management is equipped with an IP address that is assigned to it and is linked to the internet. The workstation tracks transmission parameters that are gathered from the WSN in real time. Because it is internet-connected, operations staff may access it from a distance.

10.2.2.3 Design of distribution hardware

The equipment deployment architecture for the IoT-based distribution system is shown in Figure 10.2. There are similarities between the transmission system and the WSN device, network concept, communication architecture, and data processing. But in the distribution system, keeping an eye on the substation's condition is crucial. As a result, the substation has to have a smart sensor and smart meter installed to monitor crucial

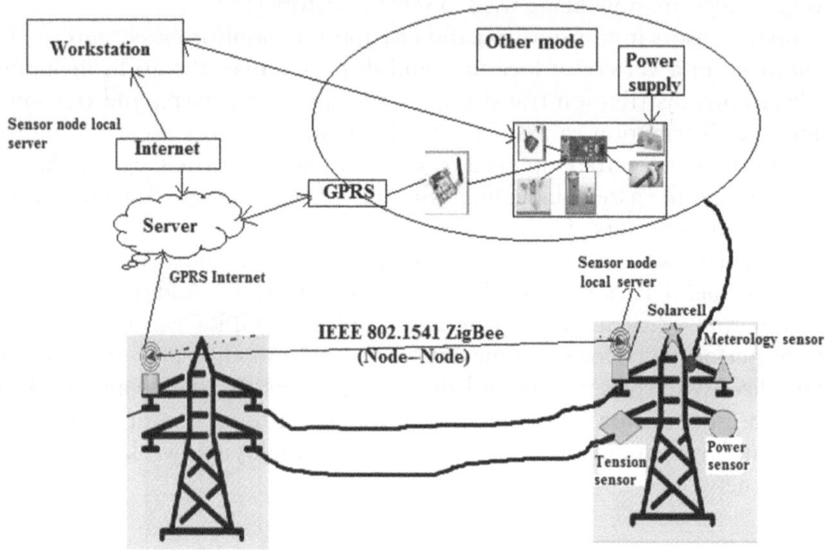

Figure 10.2 IoT-based transmission system.

parameters, including the current and voltage profiles, transformer temperature, oil level, load profiles, faults, and outages. Using cloud information management, real-time monitoring and early warning are made possible by monitoring IoT-based transmission and distribution systems. This is a potent tool for improving system dependability [11].

10.2.2.4 Design of software

Generally speaking, WSN and IoT software make up this system's software. They are both specially designed software that is deployed in sensor nodes, coordinator nodes, and embedded server, apps, and operating systems. Software flowcharts are displayed in Figure 10.3. The primary functions of the WSN software are system start-up, sensor data acquisition, and A/D conversion of data. Next, the data is transmitted to the server over GPRS or the Zigbee network.

10.3 FUTURE NEEDS FOR ADVANCED SENSORS FOR POWER SYSTEMDEVICES

- Sensors with large bandwidth and/or a broad dynamic range; those that monitor significant physical characteristics such as voltage, current, power, power factor, frequency, and harmonics;

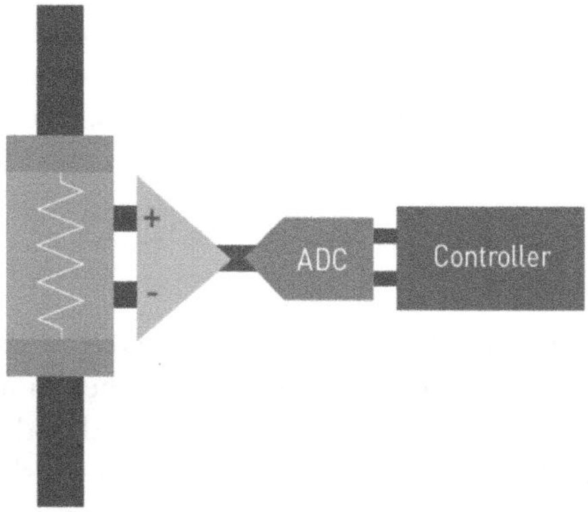

Figure 10.3 Shunt resistor usages for current sensing.

- Low-cost, high-precision sensors with high accuracy and precision;
- Self-calibrating, adaptive, and resilient sensors;
- Sensors measuring non-traditional electrical phenomena like partial discharge, electromagnetic pulses (EMPs), and geomagnetic disturbances (GMDs);
- Sensors measuring external phenomena like weather and significant human social dynamics (traffic, events);
- Self-diagnostic sensors measuring equipment health and lifetime.

10.4 CURRENT SENSOR

When electricity flows via a conductor, a current sensor finds it and measures it. It converts the current into an output that can be measured, such as a voltage, current, or digital signal, which may then be used for control, protection, or monitoring in a range of applications. To guarantee the safe and effective operation of devices and equipment, current sensors are essential parts of many electrical and electronic systems [6].

10.4.1 The benefits and drawbacks of different types of current sensors

Based on the basic physical principle that guides their design, we may classify current sensors. Ohm's Law and Faraday's Law are the modern principles for sensors, including magnetic fields, and optical sensing. Now that we have this categorization, we will discuss the several kinds of universal current sensors.

10.4.1.1 Based on Ohm's Law, current sensing

According to Ohm's Law, the voltage across a resistive device looks commensurate to the current applied across it. This basic method has been used to create a variety of sensors, such as the widely used copper lines and shunt resistors [7].

10.4.1.2 Resistor shunt

These are low-resistance precision resistors linked in series with the wire carrying the current. The current running through the shunt resistor determines the voltage drop across it, which could be quantified to determine the current passing by it. Both AC and DCs may be measured with shunt resistors. Shunt resistors can result in a significant loss of power that rises in proportion to the current squared as they are in the current-conducting path [9].

10.4.1.3 Copper trace

Alternatively, a specified shunt resistor can be used in place of the conducting element's inherent resistance inside the circuit. A high-gain parametric amplifier is necessary to provide a meaningful signal, though, because copper lines have low resistance and, thus, low voltage drops. Figure 10.3 shows an example of how a shunt resistor is commonly used for current sensing. Before being processed by a digital controller, the generated voltage is amplified to utilize the full measurement range of the analog-to-digital converter. The intrinsic electrical link between the sensing circuit and the current to be calculated is a major drawback of this type of current sensor [1].

10.4.1.4 The law of Faraday concerning current sensing

Current measurement devices frequently make use of Faraday's Law of Induction, which establishes a relationship between the total electromotive force (EMF) created in a closed circuit and the temporal rate of change of the total magnetic flux connecting the circuit. Rogowski coils and current transformers (CTs) are two popular Faraday's Law-based sensing devices. These sensors offer the necessary electrical isolation between the output signal and the current to be monitored automatically when it is essential for safety purposes. This increases the value of the current-sensing apparatus [8–10].

10.4.1.5 Transformations of current

Three windings make up the CT: a core, a secondary winding, and the main winding, which is often single-looped. The sensor is efficient at

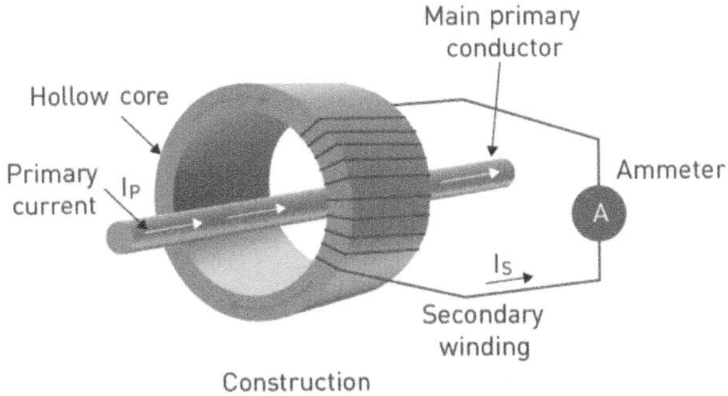

Figure 10.4 Foundational CT constructions.

sensing strong alternating currents (ACs).This allows for the transformation of a big primary current into a smaller secondary current. This device is passive, meaning it does not need any additional driving circuitry. Its ability to monitor extremely high current with very low power consumption is another noteworthy characteristic [11]. The signal will be distorted if the very high primary current or a significant component of DC in the current saturates the ferrite material employed in the core. Another issue is that, unless it is demagnetized again, the core will exhibit hysteresis after it has possessed magnetic properties, reducing precision. In addition, they are unable to detect DCs in standard form, as the basic principle depends upon the identification of a flux variation, which corresponds to a shift in current.

The functioning of CT is shown in Figure 10.4. Depending on the turn's ratio, the fluctuation in the main current I_p is reflected as I_s on the secondary side, where it can be used for sensing. By producing an output voltage proportionate to the primary current, a shunt resistor allows one to see the output current. Low losses, isolation, a simple principle of operation, and a voltage output that does not require further amplification for current sensors are all provided by this. A direct sample of the output voltage can be obtained using an analog-to-digital (A/D) converter. The primary current drop ratio is displayed by the CT ratio. The CT class, sometimes referred to as the CT class, is a metric used to quantify the accuracy of a CT.

10.4.1.6 Rogowski coils

These pliable, air-core coils are coiled around a conductor. The fluctuating magnetic field brought on by the current in the conductor I_p induces a conductor voltage that is proportionate to dI_p/dt. Rogowski coils are mostly used for measuring ACs, particularly in high-frequency applications [12].

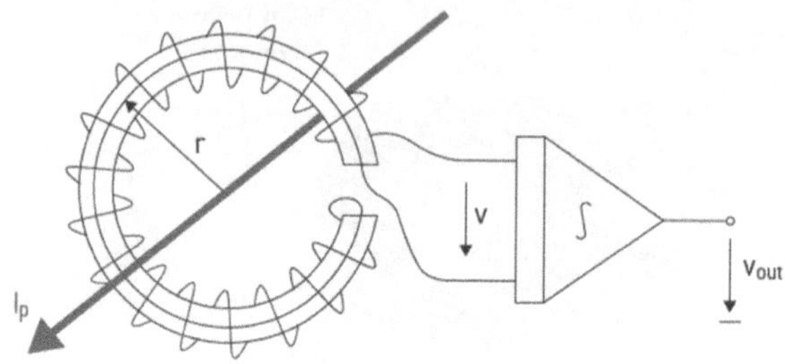

Figure 10.5 Diagram of a normal Rogowski coil.

A typical Rogowski coil's design can be seen in Figure 10.5. The voltage generated is determined by the derivative of the primary current. An integrator is needed at the output to get the necessary current sensing. Because the CT cannot use a high-permeability magnetic core, the sensitivity of the Rogowski coil is decreased.

10.4.2 Current-sensing magnetic-field sensors

The detection of currents producing static magnetic fields by using Faraday's Law is challenging. However, magnetic-field sensors can distinguish between magnetic fields that are moving and those that are static. They offer themselves as a suitable substitute option for existing sensing [13–15].The Hall-effect serves as the foundation for these sensors. The Fleming left-hand rule may be used to calculate EMF.

10.4.2.1 Sensors fluxgate

The schematic configuration of a fluxgate sensor for measuring current IP is shown in Figure 10.6. The permeability of the material changes due to the nonlinear connection between the magnetic flux density (B) and magnetic field (H), which is utilized by basic fluxgate sensors. A magnetic field is produced during excitation when the excitation winding is connected to a sinusoidal current source. Therefore, sensing may be done using the voltage that the pickup coil generates. Given that the output voltage peak is proportional to the excitation field and the external field is smaller, it may be utilized to measure the external field [16].

10.4.2.2 Sensors that are magneto-resistive

Figure 10.7 illustrates this phenomenon. The term "magneto-resistive (MR) effect" describes how the magnetic field affects material resistance. An

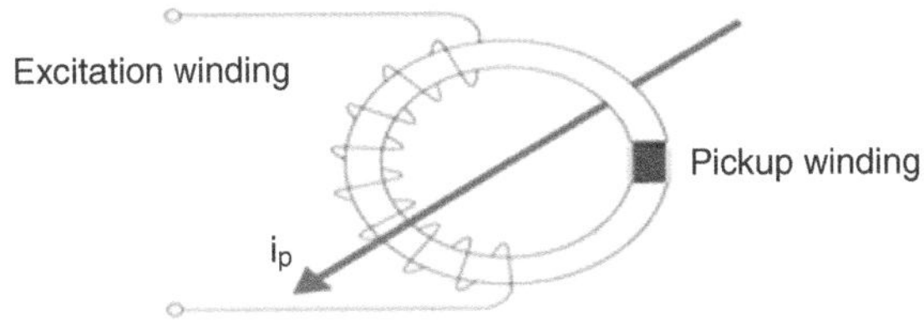

Figure 10.6 Fundamental fluxgate setup.

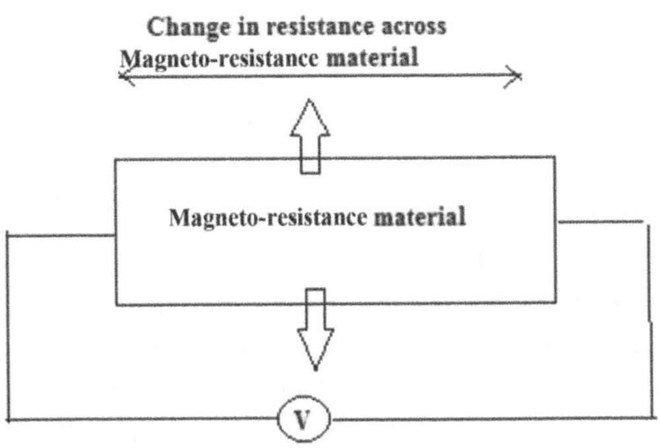

Figure 10.7 Sensory MR current principle.

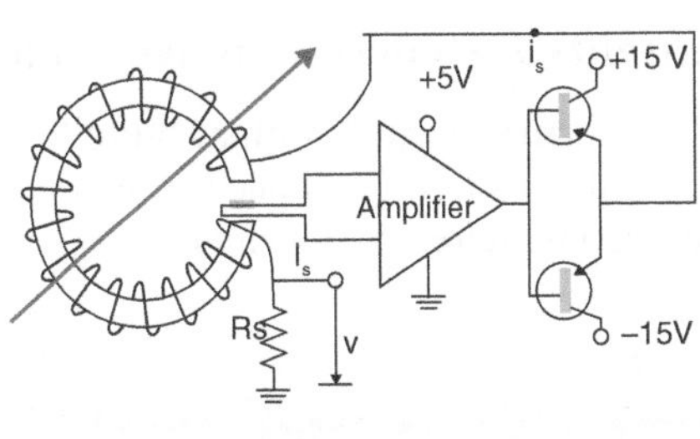

Figure 10.8 Closed and open-loop sensors experimental setups.

applied magnetic field causes a two-terminal device, a magneto-resistor, to change in resistance. By using the MR effect, differences in the magnetic field are caused by current changes to be discovered [17]. These variations are then reflected in the voltage, which may be examined further.

10.4.3 Configurations of current sensors based on magnetic fields

Figure 10.8 represents the closed and open-loop sensor experimental setup. These current sensors that rely on magnetic fields are extensively employed in both closed-loop and open-loop settings. In a closed-loop system, the current is driven via a second transformer winding using the output voltage of the magnetic-field sensor as an error signal. By creating a magnetic field that pushes the magnetic core's internal flux to zero, this current makes up for it [18–20]. The secondary current should be proportionate to the primary current in an ideal situation.

Accurate readings from open-loop sensors can be impacted by nonlinearities such as temperature drift. Closed-loop sensors, on the other hand, improve saturation and accuracy. The demands and particular applications dictate the configuration of sensors to be utilized. Open-loop current sensors are widely used as they are inexpensive, small, and have low power consumption.

10.4.4 Current sensors using fiber optics

Figure 10.9 shows how polarization might result from a magnetic field due to Faraday's effect, which could be produced by a detected current. Utilizing the Faraday magneto-optic effect, current sensors may be produced. Polarization rotation along the light transmission path is proportional to the magnetic-field projection due to the Faraday effect. This phenomenon is seen in most optically transparent dielectric materials under magnetic fields. Antennas and polarizers at the input and output may identify rotation. One may estimate the magnetic field and, by extension, current, by measuring the rotation of the incident polarization.

10.4.5 Requirements for current sensor selection

To ensure optimal performance and compatibility, a current sensor for a definite purpose, several considerations need to be taken into account. The following are some crucial factors to consider.

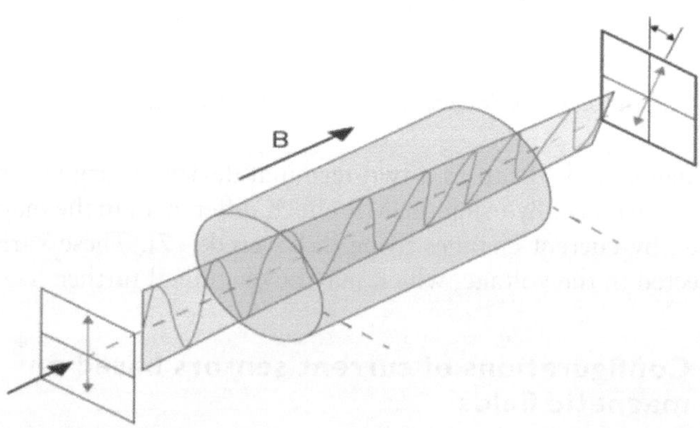

Figure 10.9 Light is polarized due to Faraday's effect.

10.4.5.1 Current type

Determine if you require measurements of DC and ACs or both types of current. While certain sensors, like shunt resistors and CTs, are only meant to monitor AC currents, others, like Hall-effect sensors, are capable of detecting both DC and ACs.

10.4.5.2 Current array

Take into account measuring the lowest and highest values of the current. Choose a sensor that suits your demands by choosing one with a suitable measuring range. Preventing saturation of the sensor and guaranteeing precise measurements over the whole range are crucial.

10.4.5.3 Precision

Analyze the precision requirements for your application. Sensitivity varies across various sensors. A few sensors could have lower precision in certain situations, but fluxgate, optical current, and precision shunt resistors all give outstanding precision.

10.4.5.4 Sensitivity

The smallest change in current that has to be recognized is the measure of sensitivity. While certain sensors are better suited for applications requiring extreme precision, others may not be sensitive enough to detect even the smallest variations in current.

10.4.5.5 Band width

Depending on the use of the current, you may need to measure it at a few specific frequencies or over a wide range. Select a sensor whose frequency response matches your requirements, especially if you are dealing with high-frequency currents.

10.4.5.6 Environmental variables

Consider the temperature range at which the sensor operates, the humidity level, and any other outside circumstances that could affect its functionality. Because certain sensors are more resilient to outside influences than others, it is important to choose one that will perform consistently under the particular needs of your application.

10.4.5.7 Dimensions and form factor

Take into account the physical constraints of the project, including mounting options and available space. One excellent illustration of a tiny, easy-to-install sensor is a shunt resistor. On the other hand, CTs might be larger and need greater space.

10.4.5.8 Need for power supply

It is significant to consider the maximum working voltage of the sensors. Ascertain applications could have specific power requirements, it is important to evaluate the losses of the power incurred by the sensors due to their load resistance.

10.5 APPLICATION FOR ADVANCED CURRENT SENSORS IN POWER SYSTEM

- *Grid monitoring and power systems:* To maintain the strength and monitor or track the transformer's currents, short circuits, and power transmission lines.
- *Over current protection:* Among other protection devices, circuit breakers, fuses, and relays use current sensors to identify overcurrent occurrences and start shielding performance, preventing harm to apparatus and guaranteeing protection.
- *Power management:* Demand-side management applications, smart meters, and energy monitoring systems employ current sensors to evaluate and control power use in residential, industrial settings, and commercial.
- *Motor control systems:* Employ current sensors to observe and regulate the current supplied to motors, enabling precise position and speed.
- *Renewably powered systems:* The solar cells and other devices' operation that generate sustainable energy require the monitoring and control of current sensors to maximize effectiveness and prevent harm from short circuits or overloading.
- *Battery management system:* To optimize performance and extend the lifetime of battery in battery management systems, current sensors are used for grid energy storage, electric vehicles, and other applications. They monitor and control charging and discharging currents.
- *Monitoring of the power supply:* To ensure dependable and efficient operation, to monitor and adjust the provided current, power supply systems for appliances and electronics use current sensors.
- *Automation and control in the industrial region:* Current sensors are used in robotics, developed, process control, and other industries

to sense and manage currents in motors, actuators, and other components.

- *Test and measurement tools:* Current sensors are used by oscilloscopes, multimeters, and other test and measurement tools to detect and analyze the electrical circuits and systems' currents.

10.6 CONCLUSION

This chapter provides a summary of the most recent advancements in sensing technology for distribution and transmission networks. The study researched, compared, and summarized the most recent literature about advanced sensors and their need on important factors of power systems. A timetable for the advanced technologies of sensing in transmission and distribution systems is also included in the study. The report also provides perspectives on sensor technologies from the utility and consumer grids. In conclusion, the study outlines the advantages of sensor technology for power systems, its potential consequences, and some of the major problems that may be of interest to researchers in the future. From the overview above, it is clear that sensing technologies are quickly growing in importance as a part of grid management and monitoring systems.

REFERENCES

1. Fe Almeida, L.E.F.; Dos Santos, J.R.; Pereira, N.A.M.; Sodre, A.C.; Mendes, L.L.; Rodrigues, J.J.P.C.; Rabelo, R.A.L.; Alberti, A.M. Control Networks and Smart Grid Teleprotection: Key Aspects, Technologies, Protocols, and Case-Studies. *IEEE Access*, *8*, 174049–174079, 2024.
2. Liu, J.; Zhao, Z.; Ji, J.; Hu, M. Research and Application of Wireless Sensor Network Technology in Power Transmission and Distribution System. *Intelligent and Converged Networks*, *1*, 199–220, 2023.
3. Pong, P.W.; Annaswamy, A.M.; Kroposki, B.; Zhang, Y.; Rajagopal, R.; Zussman, G.; Poor, H.V. Cyber-Enabled Grids: Shaping Future Energy Systems. *Advances in Applied Energy*, *1*, 100003, 2021.
4. Singh, N. "Application of Fiber Optics for the Protection and Control of Power Systems". In *Optical to Terahertz Engineering. Progress in Optical Science and Photonics*, Springer, Singapore, vol. 23, pp. 101–120, 2023. https://doi.org/10.1007/978-981-99-0228-6_7
5. Nighoskar, A.; Chaurasia, P.; Singh, N. Advancing Nuclear Energy Forecasting: Exploring Regression Modeling Techniques for Improved Accuracy. *Nuclear Engineering and Technology*, *57*(1), 103144, 2025, https://doi.org/10.1016/j.net.2024.08.013
6. Huo, Y.; Prasad, G.; Lampe, L.; Leung, V.C.M. "Advanced Smart Grid Monitoring: Intelligent Cable Diagnostics using Neural Networks". In *2020 IEEE International Symposium on Power Line Communications and its Applications (ISPLC) 2020 May 11* (pp. 1-6). IEEE; pp. 1–6, 2020.

7. Veena, R.S.; Khambra, G.; Sayal, R.; Singh, N.; Bala Dhandayuthapani, V.; Kumar, S. An Energy Efficient IPv6 Packet Delivery Outline for Industrial IoT Over G.9959 Protocol Based Wireless Sensor Network (WSN). *SN Computer Science*, *5*, 813, 2024, https://doi.org/10.1007/s42979-024-03195-4
8. Singh, N.; Sasirekha, S.P.; Dhakne, A.; Sai Thrinath, B.V.; Ramya, D.; Thiagarajan, R. IoT enabled Hybrid Model with Learning Ability for E-Health Care Systems. *Measurement: Sensors*, *24*, 1–14, 2022. https://doi.org/10.1016/j.measen.2022.100567.
9. Singh, N.; Kumar Prajapati, G. Wireless Sensor Network Security Problems—A Review. *International Journal of Emerging Technology and Advanced Engineering*, *8*(1), 2018.
10. Sondhiya, V.; Dwivedi, K.; Kulsum Almas, S.; Singh, N. "Energy Management of Nanogrid through Flair of Deep Learning from IoT Environments". In *Artificial Intelligence Techniques in Power Systems Operations and Analysis*, Taylor& Francis, 2023.
11. Wu, T.; Liu, G.; Fu, S.; Xing, F. Recent Progress of Fiber-Optic Sensors for the Structural Health Monitoring of Civil Infra-Structure. *Sensors*, *20*, 4517, 2020.
12. Xie, K.; Zhao, Y.; Zhang, H.; Zhao, Z.; Lu, Z.; Chai, Q.; Zhang, J. "Practice of Optical Fiber Sensing Technologies in Power Transmission Lines and Towers". In *Proceedings of the 2018 International Conference on Power System Technology (POWERCON), Guangzhou, China, 6–8 November*; pp. 3912–3918, 2018. IEEE
13. Hamed, M.S.; Dadashi, H.I. Survey of Different Sensors Employed for the Power Transformer Monitoring. *IET Science, Measurement & Technology*, *14*, 1–8, 2020.
14. Singh, N.; Sharma, A.K.; Sarkar, I.; Prabhu, S.; Chadaga, K. IoT-Based Greenhouse Technologies for Enhanced Crop Production: A Comprehensive Study of Monitoring, Control, and Communication Techniques. *Systems Science & Control Engineering*, *12*(1), 2024. https://doi.org/10.1080/21642583.2024.2306825.
15. Bustamante, S.; Mario, M.; Alberto, A.; Pablo, C.; Alberto, L.; Raquel, M. Dissolved Gas Analysis Equipment for Online Monitoring of Transformer Oil: A Review. *Sensors*, *19*, 4057, 2019.
16. Singh, N.; Mishra, A.; Gupta, M.; Dwivedi, K.; Swarnkar, M. "Implementation of Microgrid for Rural Electrification to Cater on Side Load Demand". In *2023 3rd International Conference on Energy, Power and Electrical Engineering (EPEE), Wuhan, China*, pp. 1058–1061, 2023. IEEE. doi: 10.1109/EPEE59859.2023.10351810.
17. Tran, Q.T.; Davies, K.; Roose, L.; Wiriyakitikun, P.; Janjampop, J.; Sanseverino, E.R.; Zizzo, G. A Review of Health Assessment Techniques for Distribution Transformers in Smart Distribution Grids. *Applied Sciences*, *10*, 8115, 2020.
18. Singh, N.; Tamrakar, S.; Mewada, A.; Gupta, S.K. *Artificial Intelligence Techniques in Power Systems Operations and Analysis* (1st ed.), Auerbach Publications, pp. 1–72, 2023. https://doi.org/10.1201/9781003301820

19. Zhao, L.; Zhang, T.; Huang, X.; Zhang, Y.; Liu, W. A Structural Health Monitoring System of the Overhead Transmission Line Conductor. *IET Science, Measurement & Technology*, *16*, 28–39, 2022.
20. Zainuddin, N.M.; Rahman, A.; Kadir, A.; Ali, N.; Ali, N.H.; Osman, Z.; Mansor, M.; Ariffin, M.M.; Rahman, A.S.A.; Nasir, M. Review of Thermal Stress and Condition Monitoring Technologies for Overhead Transmission Lines: Issues and Challenges. *IEEE Access*, *8*, 120053–120081, 2020.

Chapter 11

Navigating demand-side management in electrical power systems

A comprehensive guide to energy efficiency and grid stability

Kiran Kumari, Rishi Dev, and Liza Mozgunova

11.1 INTRODUCTION

The never-ending electricity usage around the world, coupled with the imperative to transition to sustainable energy sources, presents significant challenges and opportunities for power systems world wide. Demand-side management (DSM) emerges as a critical strategy in this context, aimed at optimizing energy consumption patterns, enhancing grid stability, and improving energy efficiency [1]. This chapter delves into the concept of DSM, exploring its techniques, benefits, challenges, and the role of technological innovations in driving it simple mutation.

11.2 DEFINITIONS AND OBJECTIVES OF DSM

DSM refers to the arrangement of measures designed to manage and modify consumer demand for electricity. It is a critical component of modern electrical power systems. It encompasses variables strategies and techniques aimed at optimizing the consumption of electricity by focusing on the consumer side [2]. This chapter encapsulates the process of DSM right from energy generation to its transmission and distribution to ultimate consumption as depicted in Figure 11.1. The chapter further explores the role of DSM in achieving energy efficiency, grid stability, and cost-effectiveness, providing insights into how it can transform the landscape of energy consumption and distribution.

DSM has evolved significantly over the decades, transitioning from basic energy conservation initiatives to sophisticated, technology-driven programs enabled by smart grids [3]. This evolution reflects the changing dynamics of the energy sector, driven by technological advancements, environmental concerns, and shifting regulatory landscapes [4].

Bangalore Electricity Supply Company Limited (BESCOM), India's major electricity distribution company, initiated its first DSM activity on November

DOI: 10.1201/9781003516156-11

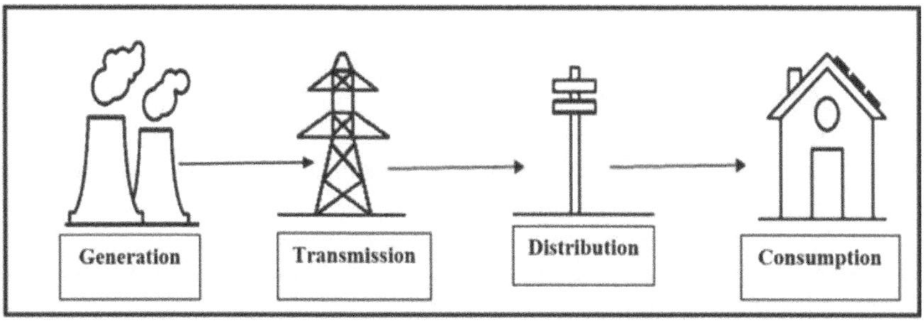

Figure 11.1 Demand-side management (DSM).

5, 2007. This marked a significant milestone for the Indian energy sector, as DSM focuses on managing and reducing electricity demand adopting tactics, such as efficient energy, load shifting, and demand response [5].

By starting DSM activities, BESCOM set an example for other electricity distribution companies in India to follow, emphasizing the importance of efficient energy management that is conducive in a more sustainable power system.

According to BESCOM,

> Demand Side Management is a utility program aiming to fine-tune consumer's energy consumption pattern according to the utility's energy demand and distribution capacity without compromising consumer's comfort. It relies on a combination of using high efficiency equipment and efficient use of electricity by the implementation of policies and measures which serve to control, influence and generally reduce electricity demand.[6]

DSM includes the planning, implementing, and monitoring of utility activities designed to encourage customers to modify their patterns of electricity usage. These programs include load management, demand response, and energy efficiency, aiming to improve energy system reliability and cutting costs [7].

> Programs initiated by utilities or other entities that aim to encourage consumers to use electricity more efficiently or shift their usage to different times of the day to balance supply and demand. These programs are keys to reducing peak loads, integrating renewable energy sources, and minimizing environmental impacts. [8]

11.2.1 Objectives of DSM

- *Reduce peak demand:* Lowering the highest level of energy demand during specific periods to prevent overloading the grid.
- *Shift load:* Moving energy consumption from peak to off-peak periods to balance the load on the grid.
- *Enhance energy efficiency:* Implementing measures that reduce total energy usage of energy resources via sustainable practices.
- *Integrate renewable energy:* Facilitating consumption of resources which are renewable in nature by aligning consumption patterns with their availability.

11.3 DSM CLASSIFICATION

DSM may be categorized based on different criteria. Here are the primary classifications:

- *Energy management programs:* A program designed to incorporate various energy and cost-saving practices. It applies various quantitative tools for tracking energy production and consumption, which further aids manufacturers in decision-making [9].
- *Demand response programs:* These programs are designed in such a way that the consumers are given an opportunity to shift energy consumption in peak hours. Their role is significant as their data are used by manufacturers and distributors to balance the supply of and demand for electricity [10].

11.4 DSM KEY TECHNIQUES

DSM employs various techniques to achieve its objectives, including:

- *Load shifting:* Encouraging consumers to use electricity during off-peak times through incentives or pricing schemes.
- *Peak shaving:* Reducing the highest levels of demand by managing or curtailing non-essential loads during peak periods.
- *Demand response:* Enabling consumers to adjust their energy use in response to real-time signals from the grid, such as price changes or grid stability notifications.
- *Energy efficiency programs:* Promoting the adoption of energy-efficient appliances, lighting, and industrial processes.

11.4.1 DSM strategies and techniques

Analysis of key DSM strategies includes the load shifting, peak shaving, energy storage, demand response, and energy efficiency programs.

11.4.1.1 A load shifting strategy

It is a pictorial illustration illustrating how electrical demand is moved from peak times to off-peak times to optimize energy use and costs. Load shifting is a DSM strategy used in energy systems to improve efficiency and reduce the necessity of added power generating capacity [9]. Here's load shifting diagram is graphically represented in Figure 11.2 that is based on Table 11.1 consisting of baseline load curve and Table 11.2 showing the load shifting plan. Both tables have taken the variables of time and energy demand into consideration, whereby the latter takes the shift of 100 units of energy for peak and off-peak hours.

The graph in Figure 11.2 helps in understanding how load shifting can help in balancing energy demand more efficiently throughout the day.

11.4.1.2 Peak shaving strategy

This is a method to reduce the peak demand of electricity in a power system by temporarily reducing or delaying certain loads during peak demand

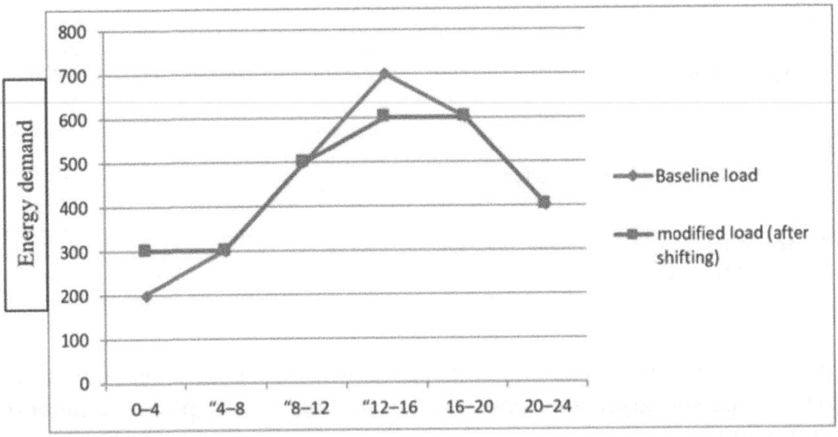

Figure 11.2 Load shifting graph.

Table 11.1 Baseline load curve

Time(hours)	0–4	4–8	8–12	12–16	16–20	20–24
Energy demand	200	300	500	700	600	400

Table 11.2 Load shifting plan [Shift 100 units from12–16 (peak) to 0–4 (off-peak)]

Time(hours)	0–4	4–8	8–12	12–16	16–20	20–24
Energy demand	300	300	500	600	600	400

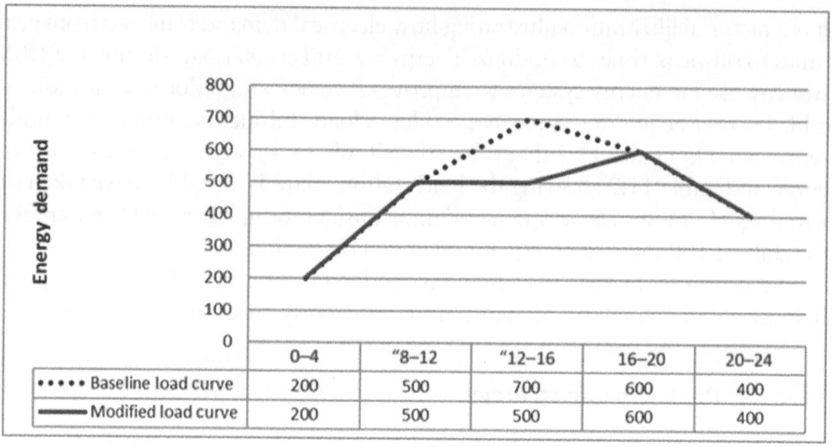

Figure 11.3 Peak shaving graph.

Table 11.3 Baseline load curve

Time(hours)	0–4	4–8	8–12	12–16	16–20	20–24
Energy demand	200	300	500	700	600	400

Table 11.4 Modified load curve for peak shaving plan [Reduce 200 units from 12 to 16 (peak)]

Time(hours)	0–4	4–8	8–12	12–16	16–20	20–24
Energy demand	200	300	500	500	600	400

periods [10]. This helps in minimizing the demand charges, reducing strain on the power grid, and potentially avoiding excess power generation dimensions

Here's a basic guide on understanding the peak shaving plan through the graphical representation in Figure 11.3, which is the outcome of the variables of time and energy demand to determine the base line curve in Table 11.3 and the variables of time and energy demand with reduction of 200 units from 12 to 16 (peak hours) in Table 11.4, to determine the modified load curve for peak shaving plan.

11.5 DEMAND-SIDE RESPONSE IMPLEMENTER

A demand-side response (DSR) implementer manages the integration of energy demand reduction or shifting strategies within power systems. They coordinate with consumers to adjust electricity usage during peak periods, aiding grid stability, and efficiency [11]. DSR implementers deploy technologies and programs to optimize energy consumption, benefiting both consumers and the grid.

11.6 CONSUMPTION OF ENERGY RESOURCES

During the FY: 2022–23, India has experienced a rather steady growth rate of consumption. Surmounting over the grappling moments of the global pandemic, all the sectors have experienced a healthy growth rate demonstrating a steady recovery of the Indian Economy [12]. We can see a steady trend in installed electricity capacity generation in megawatts between the periods of 2013 and 2023 in India, as shown in Figure 11.4.

According to the IEA, India has been an association country since March 2017. Total energy consumption(TEC) in an economy serves as a good indicator of efficient or non-efficient end use in economic activities and may indicate course correction measures to sustainability [13]. The TEC in billion kWh during the year 2022–23 with respect to its preceding years can be seen in Figure 11.5.

TEC is defined as including the total usage by the end-user segments taking into consideration the use and non-use of resources. Consumption of energy in transformers and industrial applications is not taken into consideration.

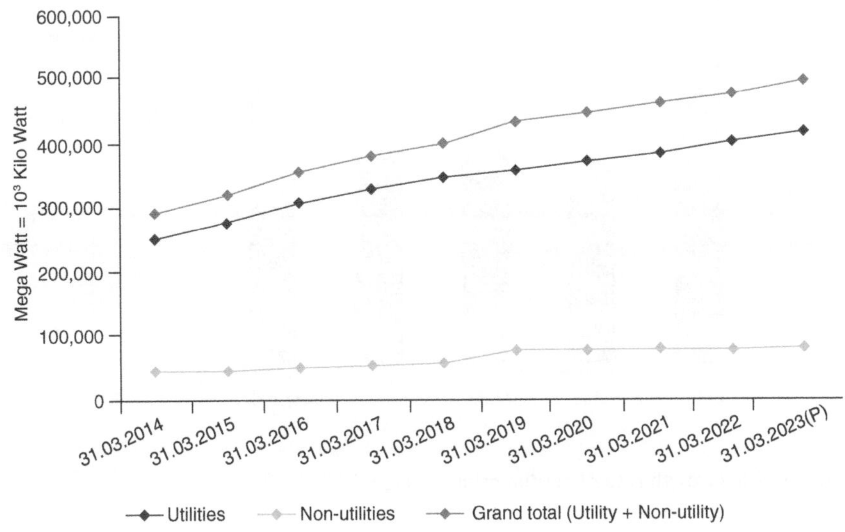

Figure 11.4 Trends in installed electricity generation capacity (MW) in India during the period2013–14to 2022–23 (P).

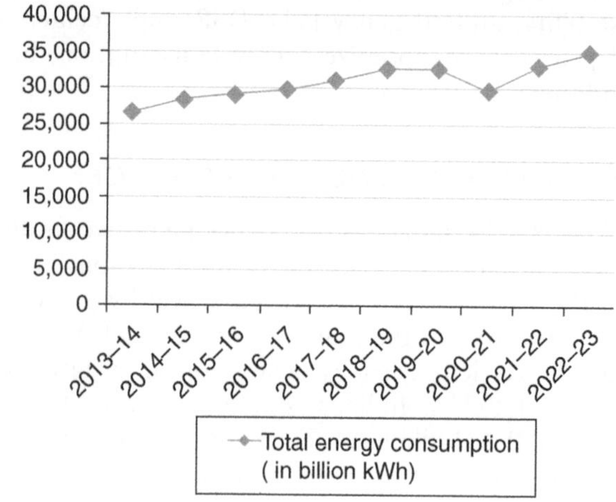

Figure 11.5 Trend of total consumption of energy in India FY:2013–14 to FY: 2022–23.

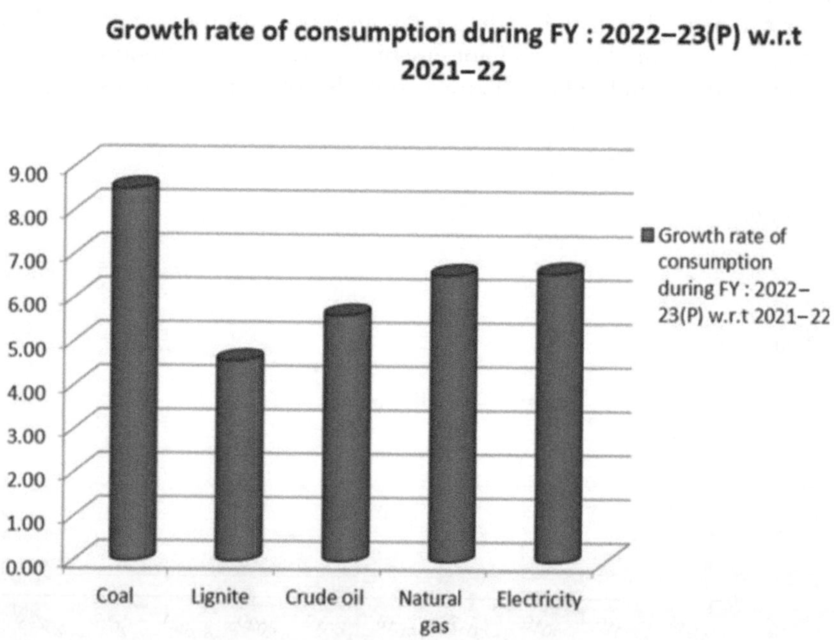

Figure 11.6 Growth rate of consumption during FY: 2022–23.

Thus, in conclusion, usage replicates the distributions to users and represents the quantity of all energy necessary to meet land consumption [14].

The TEC has increased from 33, 018PJin2021–22 to 35,159PJ in2022–23(P), an increase of 6.48%. Coal and lignite together have registered the highest growth of close to 8.63% during FY: 2022–23(P) over last year (Figure 11.6) [15].

11.7 TECHNOLOGICAL INNOVATIONS DRIVING DSM

11.7.1 Smart grid

Smart grids integrate advanced communication and control technologies to enable real-time monitoring and management of energy consumption [16]. They facilitate the implementation of DSM by providing utilities and consumers with detailed information on energy use patterns right from its inception to its end use as seen in Figure 11.7. Smart grid is a next-generation grid which is responsible for changing the lifestyle of modern society. It avoids the shortcomings of traditional grids by incorporating new technologies in the existing grids.

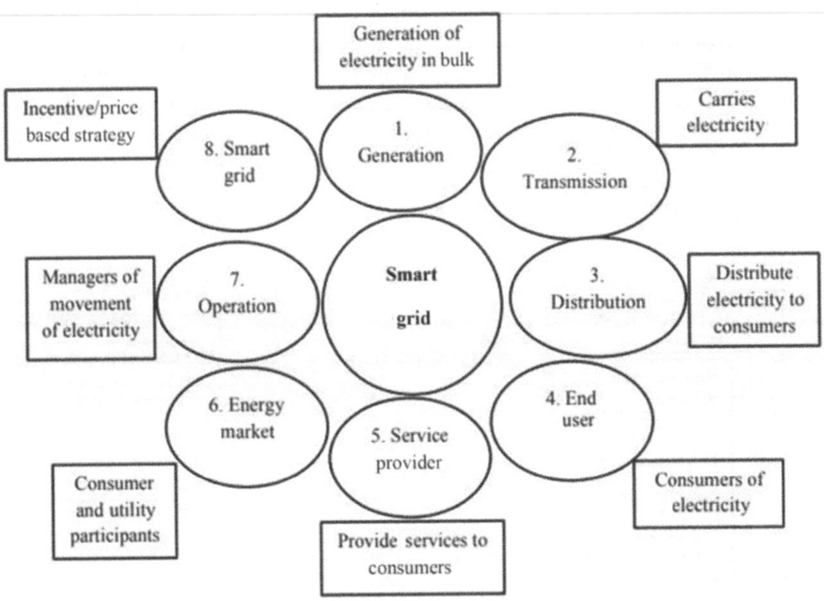

Figure 11.7 Smart grid.

11.7.2 Internet of Things (IoT)

IoT devices, such as smart thermostats and appliances, allow for automated and optimized energy use. These devices can respond to signals from the grid to adjust energy consumption based on real-time conditions.

11.7.3 Artificial intelligence and machine learning

Artificial intelligence and machine learning algorithms analyze an enormous volume of statistics to predict energy demand patterns and optimize DSM strategies [9]. They enable more precise and effective management of energy resources.

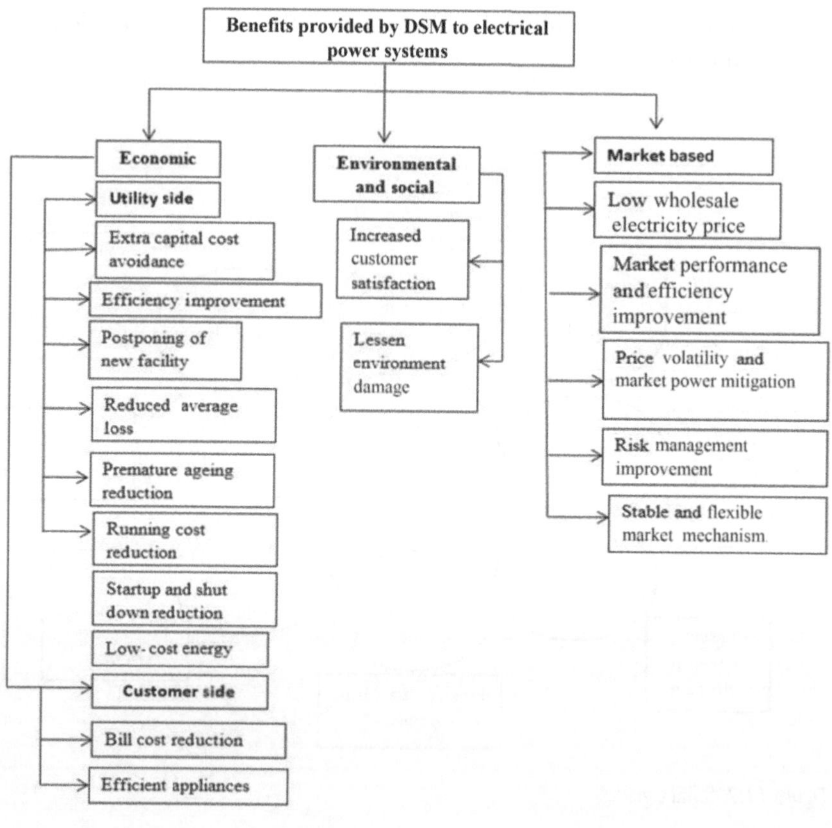

Figure 11.8 Benefit achieved by the DSM program.

11.8 BENEFIT ACHIEVED BY THE DSM PROGRAM

The benefits achieved by the DSM program are many fold. There are economic benefits for both utility side and the customer side. The DSM also benefits the environmental and social stake holders by lessening the environmental damage and increasing customer satisfaction. The benefits derived from the DSM program can be seen in Figure 11.8.

11.9 ELECTRICITY REGULATORY COMMISSION (DSM) REGULATIONS

- **Electricity Act, 2003**

Section 61: Provides for the determination of tariff by the appropriate regulatory commission, which may include provisions for promoting energy efficiency and conservation.

Section 86(1)(e): Empowers regulatory commissions to promote cogeneration and generation of electricity from renewable sources, which can contribute to DSM efforts.

- **Energy Conservation Act, 2001**

Section 14 mandates energy-intensive industries and other designated consumers to undertake energy audits and implement energy conservation measures, which can include DSM strategies.

- **National Electricity Policy**

Provides a framework for the development of the power sector in India, including objectives related to energy efficiency, DSM, and grid stability.

- **National Action Plan on Climate Change (NAPCC)**

Contains various missions and initiatives aimed at addressing climate change, including the National Mission for Enhanced Energy Efficiency, which promotes DSM and energy conservation measures.

- **Regulatory Orders and Guidelines**

Various orders and guidelines issued by regulatory commissions such as the Central Electricity Regulatory Commission (CERC) and State Electricity Regulatory Commissions (SERCs) may contain provisions related to DSM, energy efficiency, and demand response[17].

- **State Electricity Regulatory Commissions(SERCs) Regulations**

Each state may have its own regulations and orders issued by the SERC, which may include provisions related to DSM programs, energy efficiency targets, and incentives for DSM measures.

- **Renewable Energy Policies**

Various policies and schemes promoting renewable energy deployment, such as the National Solar Mission and various state-level solar and wind energy policies, indirectly support DSM efforts by diversifying the energy mix and reducing reliance on conventional power plants.

- **DSM Programs**

Utilities and government agencies may implement DSM programs and initiatives voluntarily or as part of regulatory mandates. These programs may include energy efficiency rebate programs, demand response in iterative, and consumer education campaigns.

11.10 DSM PLAN AND PROGRAMMED COMPLETION REPORT

The DSM Plan provides methods for efficiently controlling the demand for electricity, including steps to encourage energy conservation and lower peak demand. Timelines for execution, goals, and targeted industries are all included.

(a) Every 6 months, the Distribution Licensee is required to produce reports on the DSM Plan's implementation costs and progress.
(b) Within 3 months of the program's completion, the Distribution Licensee shall compile and submit a comprehensive Programmed Completion Report to the Commission.

11.11 SUSTAINABILITY AND ENERGY

By the end of third quarter of 2015, the general assembly of UN adopted the agenda 2030. It consists of more than a dozen Sustainable Development Goals (SDGs). SDGs are an all-inclusive global guideline considering the socio-economic-enviro facets of development. Resources which are sustainable and renewable are wind, water, solar, biomass, and geothermal energy is inexhaustible and clean [11].

The targets adopted as a fragment of SDG 2030 Agenda are as follows:

I. Modern energy services: Universal access, affordability, and reliability.
II. Global energy mix: Proportionate rise in renewable energy.
III. Energy efficiency: Twice as much the rate of improvement.

11.12 CONCLUSION

This chapter presents a comprehensive overview of the critical role DSM plays in enhancing energy efficiency and ensuring grid stability within electrical power systems. Smart grids are a cornerstone of modern energy infrastructure, providing substantial enhancements over outdated power grids. By enhancing efficiency, reliability, and sustainability, smart grids are transforming the consumption and management of electricity and are also playing a significant part in shaping modern society. By empowering consumers with the knowledge and tools to optimize their electricity usage, DSM initiatives can significantly contribute to mitigating peak demand, reducing strain on infrastructure, and ultimately improving the reliability and affordability of electricity services.

REFERENCES

1. Thakur, J., & Chakraborty, B., "Demand side management in developing nations: A mitigating tool for energy imbalance and peak load management," *Energy*, 114, 895–912, 2016, https://doi.org/10.1016/j.energy.2016.08.030.
2. Bakare, M. S., Abdulkarim, A., Zeeshan, M., & Shuaibu, A. N., "A comprehensive overview on demand side energy management towards smart grids: Challenges, solutions, and future direction," *Energy Informatics*, 6(1), 2023, https://doi.org/10.1186/s42162-023-00262-7.
3. Aline Ribas, A. F., "Closing the energy divide in a climate-constrained world: A focus on the buildings sector," (M. S. Jianlei Niu, Ed.),*Energy and Buildings*, 199, 264–274, 2019, https://doi.org/10.1016/j.enbuild.2019.06.053.
4. Singh, N., "Renewable energy for sustainable rural development and challenges," *IJRAR*, 6(1), 185–187, February 2019.
5. Singh, N., Chakrabarti, T., Chakrabarti, P., Margala, M., Gupta, A., Praveen, S. P., Krishnan, S. B., & Unhelkar, B. "Novel heuristic optimization technique to solve economic load dispatch and economic emission load dispatch problems," *Electronics*, 12(13), 2921, 2023. https://doi.org/10.3390/electronics12132921.
6. Singh, N., Chakrabarti, T., Chakrabarti, P., Panchenko,V., Budnikov, D., Yudaev, I., & Bolshev, V., "Analysis of heuristic optimization technique solutions for combined heat-power economic load dispatch," *Applied Sciences*, 13(18), 10380, 2023. https://doi.org/10.3390/app131810380.
7. Sondhiya, V., Dwivedi, K., Kulsum Almas,S., & Singh, N., "Energy management of nanogrid through flair of deep learning from IoT environments," *Artificial Intelligence Techniques in Power Systems Operations and Analysis*, Taylor & Francis, July 2023.
8. Bari, A., Jiang, J., Saad, W., & Jaekel, A., "Challenges in the smart grid applications: An overview," *International Journal of Distributed Sensor Networks*, 10(2), 974682, 2014, https://doi.org/10.1155/2014/974682.
9. Singh, N., Mishra, A., Gupta, M., Dwivedi, K., & Swarnkar, M., Implementation of Microgrid for Rural Electrification to Cater on Side Load Demand, *2023 3rd International Conference on Energy, Power and Electrical Engineering (EPEE), Wuhan, China*, 2023, pp. 1058–1061, https://doi.org/10.1109/EPEE59859.2023.10351810.
10. Paddy Finn, C. F., "Demand side management of industrial electricity consumption: Promoting the use of renewable energy through real-time pricing,"

(J. Yam, Ed.), *Applied Energy*, 113, 11–21, 2014, https://doi.org/10.1016/j.apenergy.2013.07.003.

11. Singh, N., Almas, S. K., Tirole, R., Jain, A., Shukla, S., & Kumar, Y., Analysis of Optimum Cost and Size of the Hybrid Power Generation System Using Optimization Technique, 2023 IEEE 12th International Conference on Communication Systems and Network Technologies (CSNT), Bhopal, India, 2023, pp. 284–291, https://doi.org/10.1109/CSNT57126.2023.10134720.
12. Singh, N., Priyadarshi, N., & Kumar, N., Meta Heuristic Techniques Used to Improve the Performance of Static Converters, *International Conference on Innovative Development in Virtual and Rapid Prototyping (IDVRP-2022)*, 14–15 May 2022, Dehradun, India. *AIP Conference Proceedings 2771, 020027*, 2023 (Scopus Indexed), https://doi.org/10.1063/5.0152913.
13. Bakare, M. S., Abdulkarim, A., Zeeshan, M., & Shuaibu, A. N., "A comprehensive overview on demand side energy management towards smart grids: challenges, solutions, and future direction," *Energy Inform* 6, 2023,
14. Singh, N., Priyadarshi, N., & Kumar, N., Minimization of Environmental Emission and Cost of Generation Using Economic Load Dispatch with Optimization Techniques. *AIP Conference Proceedings* 2771, 020025, 2023, https://doi.org/10.1063/5.0152915.
15. Singh, N., Chakrabarti, T., Chakrabarti, P., Panchenko, V., Budnikov, D., Yudaev, I., & Bolshev, V., "Analysis of heuristic optimization technique solutions for combined heat-power economic load dispatch," *Applied Sciences*, 13(18), 10380, 2023, https://doi.org/10.3390/app131810380.
16. Singh, N., Chakrabarti, T., Chakrabarti, P., Margala, M., Gupta, A., Krishnan, S. B., & Unhelkar, B., "A new PSO technique used for the optimization of multiobjective economic emission dispatch," *MDPI Electronics*, 12(13), 2960, 3 July 2023, https://doi.org/10.3390/electronics12132960.
17. Singh, N., & Kumar, Y., "Multi objective economic load dispatch problem solved by new PSO," *Advances in Electrical Engineering*, 2015, 2356–6655, 2015, https://doi.org/10.1155/2015/536040.

Chapter 12

Exploring the synergy of artificial intelligence for integrated smart energy systems in electric vehicles

A. S. Telang, G. S. Gadge, D. A. Shahakar, and Y. Pahariya

12.1 INTRODUCTION

The rise of electric vehicles (EVs) has been significantly driven by the global move toward sustainable energy solutions. EVs play a vital role in this transition by offering a more environmentally friendly alternative to traditional gasoline-powered cars, which helps reduce greenhouse gas emissions and our dependence on fossil fuels. However, the widespread adoption of EVs faces several challenges, particularly in terms of reliability, efficiency, and energy management [1].

This is where artificial intelligence (AI) steps in as a crucial technological innovation that can address these challenges and enhance the performance of EVs. AI encompasses technologies like machine learning, neural networks, and optimization algorithms, which provide sophisticated tools for managing the complex energy systems in EVs. AI algorithms facilitate real-time monitoring and control of energy storage systems, enabling EVs to adapt to changing energy demands and grid conditions. Despite its potential, this area of research is still emerging, with fewer studies compared to other fields [2].

12.2 LITERATURE REVIEW

AI has attracted a lot of attention recently, with much research examining how it might improve sustainability, dependability, and efficiency in EV energy systems. Battery management systems (BMS), energy harvesting and regeneration, route and energy optimization, smart grid interactions, predictive maintenance, and the relationship between AI, the Internet of Things (IoT), and big data analytics are just a few of the important topics that this literature review summarizes and classifies [4].

12.2.1 Battery management systems

Because battery management directly affects battery lifespan, safety, and efficiency, it is essential for EV performance. Machine learning and neural

DOI: 10.1201/9781003516156-12

networks, in particular, are two AI techniques that have been widely used in BMS to enhance state-of-charge (SoC) and state-of-health (SoH) predictions [1, 2].

12.2.2 Energy harvesting and regeneration

To increase the range and efficiency of EVs, energy harvesting and regeneration are essential. AI techniques have been used to optimize regenerative braking systems, including fuzzy logic and evolutionary algorithms. The use of fuzzy logic control to increase regenerative braking efficiency and improve energy recovery is demonstrated in a study by the authors of [3, 4] and the utilization of particle swarm optimization (PSO) and other optimization approaches to optimize the energy harvested from renewable sources integrated with EVs have been studied [5].

12.2.3 Route and energy optimization

Reducing energy consumption and increasing the range of EVs require effective route planning and energy management. Neural networks and reinforcement learning are two AI-based techniques that have demonstrated encouraging outcomes in this field. By using reinforcement learning to create a real-time energy management plan for EVs, the authors of [5] were able to save a substantial amount of energy. Furthermore, the authors of [6] have investigated neural network-based techniques that offer dynamic route optimization that takes into account current traffic and road conditions, hence improving energy efficiency even more.

12.2.4 Interactions between smart grids and EVs

An essential component of integrated smart energy systems is the interaction between smart grids and EVs. AI is essential to maximizing this relationship as it makes load balancing and energy exchange more effective. The authors of [7] and [8] also done a thorough review of the synergy between AI and energy storage systems for EVs. It investigates the importance of AI advancements in energy storage systems for EVs, specifically focusing on BMS, power quality (PQ) issues, predicting battery SoC and SoH, and exploring the potential for integrating renewable energy sources with EV charging needs and optimizing charging cycles [8].

12.2.5 Predictive maintenance

Predictive maintenance uses AI to foresee and stop possible EV component failures, improving dependability and lowering maintenance costs [9]. Support vector machines (SVMs) and random forests are two examples of machine learning algorithms that have been successfully used to forecast component failures and plan preventive maintenance. While the authors of [10] use random forests for the early detection of motor problems, Liu et al.'s [11] research emphasizes the use of SVM for battery failure prediction.

12.2.6 AI, IoT, and big data analytics

When AI, IoT, and big data analytics come together, they establish an integrated ecosystem that greatly improves the efficiency of EVs. Large volumes of data are gathered by IoT devices from different EV components, which AI systems subsequently examine to glean insightful information. Adaptive control, predictive analytics, and real-time monitoring are made possible by this combination. Research by the authors of [11] and [12] highlights how AI and IoT integration can be used to build intelligent, responsive EV systems that adjust to changing circumstances and continually improve performance.

12.3 CURRENT STATE OF ART OF EV TECHNOLOGY

The automotive industry has experienced major breakthroughs due to the rapid evolution of EV technology [12]. Modern EV technology exhibits notable advancements in car performance, charging infrastructure, and battery economy [13]. Driving ranges have been increased by advances in battery chemistry and energy density, and the convenience of owning an EV has been improved by the availability of faster and more frequent charging stations. Furthermore, energy management and vehicle control systems have been optimized through the integration of advanced AI and machine learning algorithms, increasing efficiency and dependability even further.

12.3.1 Electric powertrains

Electric powertrains are the core of EV propulsion, converting electrical energy into mechanical motion. They comprise transmission systems, inverters, and electric motors. Electric motors are crucial for efficient propulsion, as they transmit power and torque to the wheels. Inverters convert the direct current (DC) from the battery into alternating current (AC) for the motor, while gearbox systems adjust the motor's torque and speed to drive the wheels, ensuring optimal vehicle performance [14]. Refer to Figure 12.1 for the main components of an EV power train.

12.3.2 Energy storage systems

Energy storage in EVs relies on high-capacity batteries that provide electrical energy for propulsion and auxiliary systems. The predominant technology

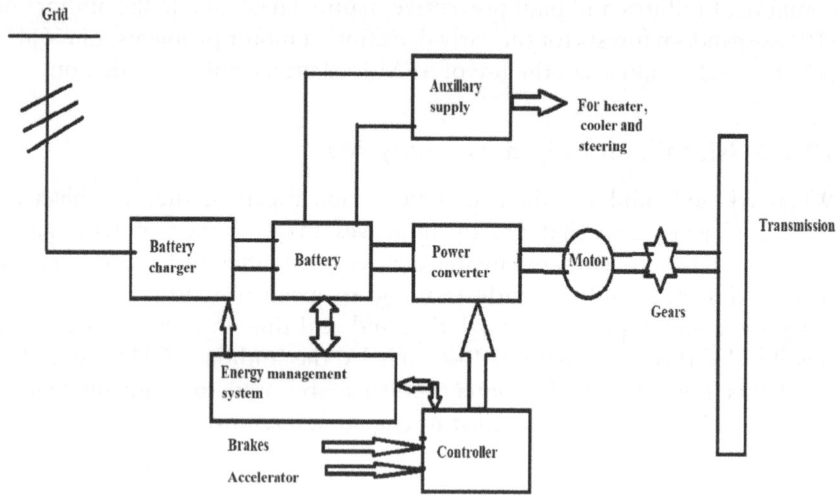

Figure 12.1 Main components of EV powertrain.

is lithium-ion batteries, known for their high energy density, long cycle life, and short charging times. A BMS regulates charging and discharging processes and monitors the state of charge to ensure both performance and safety [14].

12.3.3 EV charging infrastructure

Widespread adoption of EVs necessitates a robust charging infrastructure, comprising a network of stations from home units to public fast-charging facilities. Table 12.1 shows three levels of charging. Level 1 chargers offer low-power charging via standard home outlets, while Level 2 chargers provide faster charging at specialized stations. DC fast chargers enable quick recharging for long-distance travel. The effectiveness of these stations depends on their location, influenced by environmental, economic, and demographic factors [15]. Charging system designs should enhance consumer confidence through familiarity, simplicity, and context sensitivity. Collaboration among stakeholders is essential to address challenges and promote the broader adoption of EVs. We analyze the future of EV charging infrastructure with a focus on market forces and system solutions.

12.4 INTRODUCTION TO AI TECHNOLOGIES

AI technologies have greatly improved the creativity, efficiency, and safety of EVs. Key developments include neural networks and machine

Table 12.1 Charging levels

Charging levels		
Level 1	*Level 2*	*DC fast charge*
Voltage 120V 1-phase AC **Amps** 12–16 Amps **Charging load** 1.4–1.9 kW **Charging time** 3–5 miles/h	**Voltage** 208V or 240V 1-phase AC **Amps** 12–80 Amps **Charging load** 2.5–19.2 kW **Charging time** 12–60 miles/h	**Voltage** 208V or 480V 3-phase AC **Amps** >100 Amps **Charging load** 50–350 kW **Charging time** 10%–80% in ~30 minutes

learning algorithms, which are essential for autonomous driving, the high-performance control of electric components, and information security. AI is used in EVs to prevent attacks, detect intrusions, and ensure authentication, while also providing protection against adversarial attacks and ensuring compliance with regulations. These technologies support real-time decision-making which is crucial for the accurate operation of self-driving cars and their perception systems [16]. AI frameworks, which include neural networks and various control systems, work together to manage and monitor the power consumption of EVs, promoting safe and efficient autonomous driving.

12.4.1 The role of AI in the development and optimization of EVs

Energy conservation and environmental protection are paramount concerns in the 21st century, spurring the move towards sustainable development. This focus has driven significant advancements in EV technology, improving traffic safety and road efficiency through AI-based systems. AI applications make EVs more appealing to consumers by incorporating features such as driver assistance systems, autonomous driving, and optimized charging processes [17]. They also enhance energy management, battery performance, predictive maintenance, intelligent charging infrastructure, and integration with the smart grid. As AI continues to evolve within the EV sector, there is great potential for further performance enhancements and increased sustainability. The role of AI in the development and optimization of EVs is enlisted as follows:

1. *Maximizing vehicle efficiency through AI:* AI plays a crucial role in enhancing the efficiency of EVs by optimizing energy usage. By analyzing vast amounts of data from sensors, batteries, and various

vehicle components, AI algorithms can make real-time adjustments that greatly improve energy consumption. These algorithms are particularly effective in optimizing powertrains, managing regenerative braking, and predicting energy requirements based on driving habits, weather conditions, and traffic patterns.

2. *Autonomous driving and AI integration in EVs:* AI plays a vital role in the development of autonomous driving for EVs. AI algorithms process data from sensors such as cameras, radar, and lidar to understand the vehicle's environment and make informed decisions. This integration not only improves safety and reduces the risk of accidents but also adds convenience by taking over driving tasks.
3. *Predictive maintenance empowered by AI:* AI is transforming maintenance routines in EVs with predictive algorithms. These algorithms analyze real-time data from various components to identify potential issues before they lead to failures. By spotting anomalies and predicting maintenance needs, AI allows for proactive maintenance, reducing downtime and enhancing vehicle performance.
4. *AI-powered charging infrastructure advancements:* AI greatly enhances the efficiency and reliability of charging infrastructure for EVs. It helps by optimizing the placement of charging stations, predicting demand trends, and streamlining the charging process, which reduces wait times and improves the overall charging experience. Additionally, AI facilitates vehicle-to-grid communication, enabling EVs to act as energy storage units and help balance the energy supply.
5. *Enhancing battery technology with AI:* Advanced battery technologies are paramount for widespread EV adoption, and AI plays a critical role in enhancing battery performance, lifespan, and safety. AI algorithms monitor battery health, predict degradation, and optimize charging cycles, maximizing battery efficiency.
6. *AI's role in energy management for EVs:* Effective energy management is crucial for EVs, and AI provides smart strategies to optimize energy use. By analyzing data from the vehicle, the grid, and external sources, AI algorithms create efficient plans for charging and discharging. This helps ensure that EVs make the most of renewable energy sources, cut down on energy waste, and lower overall carbon emissions.
7. *AI-enhanced user experience in EVs:* AI enhances user experience in EVs by offering personalized features and user-friendly interfaces. Natural language processing allows for voice commands and control, while machine learning algorithms adapt to individual preferences by adjusting vehicle settings. AI-powered infotainment systems provide real-time information, entertainment options, and advanced driver assistance features.

12.4.2 Enhancing BMS in smart energy systems for EVs with AI: A case study

The rapid rise of EVs has underscored the need for advanced and effective BMS to enhance efficiency and extend battery life. As EV adoption grows, integrated smart energy systems are crucial for ensuring that these vehicles operate efficiently within the broader energy grid. Managing the complexities of energy distribution, charging requirements, and battery health monitoring requires sophisticated solutions to handle this integration effectively.

In AI, one of the major challenges is finding the best solution within a given search space, a common issue in areas like machine learning, deep learning, knowledge representation, reasoning, and perception. Traditional search methods often use heuristics tailored to specific problems, focusing on finding good enough solutions rather than perfect ones. Recently, metaheuristic algorithms have gained popularity in the AI research community due to their impressive ability to navigate complex optimization problems. These algorithms offer a flexible framework for solving a variety of optimization issues by using iterative evolution strategies. They combine learning mechanisms to make the most of accumulated knowledge while exploring the search space.

AI-based methods offer promising opportunities to enhance BMS within interconnected energy systems. AI enables real-time optimization, adaptive control, and advanced predictive analytics that surpass the capabilities of traditional methods. This case study explores how AI can be used to improve BMS in integrated smart energy systems for EVs. The findings aim to demonstrate how transformative AI can be in creating an effective, robust, and user-friendly ecosystem for EVs. By integrating AI with energy systems and battery management, we can advance towards a more advanced and sustainable future in transportation.

Using AI-based optimization algorithms to manage the charging and discharging cycles in BMS for EVs can greatly enhance energy efficiency and extend battery life. This section details the results of applying different optimization algorithms, backed by mathematical modeling, to achieve these improvements.

12.5 MATHEMATICAL MODELS

12.5.1 Battery lifespan (in number of cycles)

The battery lifespan in a number of cycles can be modeled based on factors like depth of discharge (DoD), temperature, charge/discharge rates, and other operational conditions. A common approach is to use empirical models derived from experimental data:

$$\text{Lifespan} = \frac{C_{\text{rated}}}{\alpha_O D_O D + \beta_T + \gamma C_r} \tag{12.1}$$

where:

- C_{rated} is the rated capacity of the battery.
- DoD is the depth of discharge.
- T is the operating temperature (in Celsius).
- C_r is the charge/discharge rate.
- alpha, beta, and gamma are coefficients determined through regression analysis or experimental data fitting.

12.5.2 Energy efficiency (in %)

Energy efficiency is defined as the ratio of the energy output during discharge to the energy input during charge. It is influenced by factors such as internal resistance, charge/discharge rates, and temperature. Mathematical Model:

$$\text{Efficiency } \% = \left(1 - \frac{R_{\text{int}} \times I^2}{V_{\text{avg}}}\right) \times 100 \tag{12.2}$$

where:

- R_{int} is the internal resistance of the battery.
- I is the current during charge/discharge.
- V_{avg} is the average voltage during charge/discharge.

12.5.3 Average charge/discharge time

The average charge/discharge time depends on the capacity of the battery, the charge/discharge rate, and the efficiency of the charging system.

Mathematical model:

$$t_{\text{avg}} = \frac{C_{\text{rated}}}{I \times \eta} \tag{12.3}$$

where:

- C_{rated} is the rated capacity of the battery (Ah).
- I is the charging/discharging current (A).
- η is the charge/discharge efficiency (expressed as a decimal, e.g., 0.9 for 90% efficiency).

12.5.4 Combined optimization model

These models can be integrated into a multi-objective optimization framework for optimization, with the goals of maximizing battery lifespan, maximizing energy efficiency, and minimizing charge/discharge time.

Objective function

Maximize

$$F(x) = w_1 \times \frac{C_{rated}}{(\alpha \times D_O D + \beta \times T + \beta \times C_r)} + w_2 \times \left(1 - \frac{R_{int} \times I^2}{V_{avg}}\right) - w_3 \times \frac{C_{rated}}{I \times \eta} \tag{12.4}$$

where

- x = (DoD, T, Cr, I) is the vector of decision variables.
- (w_1, w_2, w_3) are weights assigned to each objective based on their relative importance.

Constraints

1. $0 \leq DoD \leq 1$ (depth of discharge should be between 0% and 100%)
2. $T_{min} \leq T \leq T_{max}$ (operating temperature range)
3. $C_{rmin} \leq C_r \leq C_{rmax}$ (charge/discharge rate limits)
4. $I_{min} \leq I \leq I_{max}$ (current limits)

12.6 STANDARD DATA AND PARAMETER VALUES

Here are some typical data and parameter values that can serve as a starting point for optimizing battery lifespan, energy efficiency, and average charge and discharge times in BMS for AI-integrated smart energy systems in EVs.

1. *Battery lifespan:* The lifespan of a typical lithium-ion battery used in EVs can range from 500 to 3000 charge/discharge cycles, depending on various factors like battery chemistry, usage patterns, temperature conditions, and DoD.
2. *Energy efficiency:* The energy efficiency of a battery system is influenced by factors like charging/discharging rate, temperature, battery chemistry, and management algorithms. For lithium-ion batteries, typical efficiencies range from 80% to 95%.

3. *Average charge/discharge time:* The average charge/discharge time depends on the battery capacity, charging/discharging rate, and efficiency. A typical charge/discharge time for EVs batteries can range from 2 to 8 hours, with faster-charging technologies aiming for lower times.

12.6.1 Parameter optimization

To optimize these parameters using the algorithms, viz. Genetic Algorithm (GA), PSO, hybrid GA–PSO, JAYA, Learning Cooking Algorithm (LCA), and Alibaba and the Forty Thieves Algorithm, we need to define the parameter space and constraints for each algorithm. Here is a general approach:

1. *Define parameter space:* Specify the range of values each parameter can take.

   For example:
   - Battery lifespan: 500 to 3000 cycles
   - Energy efficiency: 80% to 95%
   - Charge/discharge time: 2 to 8 hours
2. *Objective function:* Define an objective function that evaluates the fitness of a solution based on the parameters (battery lifespan, energy efficiency, charge/discharge time). This function will be used by optimization algorithms to search for optimal solutions.
3. *Constraints:* Define any constraints that must be satisfied. For example, the total energy stored in the battery should meet the vehicle's requirements, and the charge/discharge rate should not exceed safe limits.
4. *Optimization algorithm:* Choose an optimization algorithm, viz.GA, PSO, hybrid GA–PSO, JAYA, LCA, and Alibaba Algorithm and set its parameters.
5. *Run optimization:* Run the chosen optimization algorithm with the defined objective function and constraints. The algorithm will search for parameter values that optimize the objective function while satisfying the constraints.
6. *Evaluate results:* After optimization, evaluate the results to see the optimized values for battery lifespan, energy efficiency, and charge/discharge time.

12.7 RESULTS AND DISCUSSION

This section presents the performance metrics of different optimization algorithms applied to BMS in AI-integrated smart energy systems for EVs. The evaluated algorithms include GA, PSO, hybrid GA–PSO, JAYA, LCA,

Table 12.2 Comparison of optimization algorithms

Algorithm	*Battery lifespan (cycles)*	*Energy efficiency (%)*	*Average charge/ discharge time (hours)*
For population size = 50; max iterations = 100			
GA	814.054	82.8092	0.72865
PSO	1000.9674	96.4176	1.9907
Hybrid GA–PSO	950.1728	94.2315	1.8276
JAYA	**2668.5079**	**94.94**	**2.0000**
LCA	977.2623	84.6803	1.8659
ABFTA	982.6657	91.3508	1.0657
Population size =100; max iterations =200			
GA	997.3475	53.9052	0.38828
PSO	1055.9573	101.7763	0.64117
Hybrid GA–PSO	954.628	96.7096	1.607
JAYA	**3000**	**93.6811**	**2.3013**
LCA	979.199	92.771	1.3449
ABFTA	995.0773	85.799	1.8881
Population size =300; max iterations =500			
GA	997.4759	91.8433	1.5989
PSO	1050.8594	95.066	0.37757
Hybrid GA–PSO	943.4707	94.3167	0.89798
JAYA	**2999.4513**	**95**	**2**
LCA	988.2624	97.3612	1.1734
ABFTA	987.5547	97.975	0.72034

and Artificial Bee Flower Pollination Algorithm. Table 12.2 represents a comparison of these algorithms for incremental population size for 50,100, and 300 and maximum iterations for 100,200, and 500.

Performance analysis

Figures 12.2, 12.3, and 12.4 illustrate the comparison of key parameters: battery lifespan, energy efficiency, and charge/discharge time. The comparison includes advanced optimization algorithms such as GA, PSO, hybrid GA–PSO, JAYA, LCA, and, Alibaba and forty thieves algorithm.

The results indicate that, with a similar average charge/discharge time, the JAYA algorithm excels over other algorithms in both energy efficiency and battery lifespan. Additionally, the PSO algorithm and ABFTA also showed strong performance, particularly in enhancing energy efficiency and extending battery lifespan.

Therefore, AI-integrated algorithms offer substantial improvements in battery lifespan and efficiency compared to traditional BMS solutions. Unlike conventional methods, AI's adaptive learning capabilities allow it to adjust to changing conditions and continuously optimize performance.

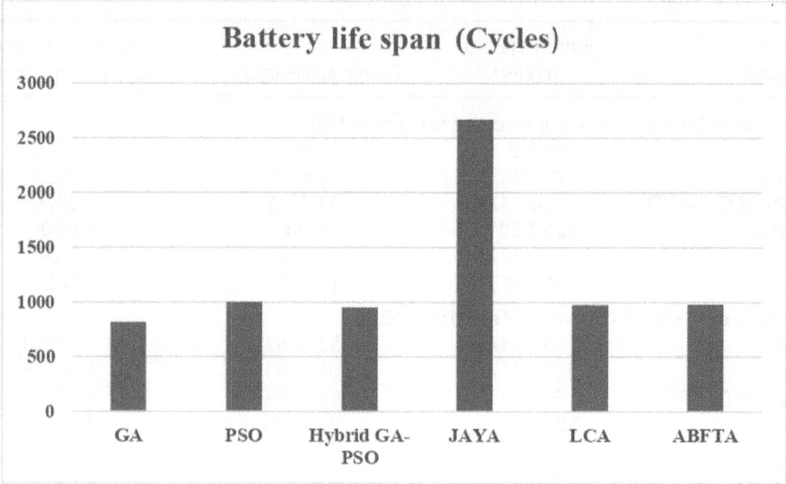

Figure 12.2 Comparison of optimization algorithm for battery lifespan parameter.

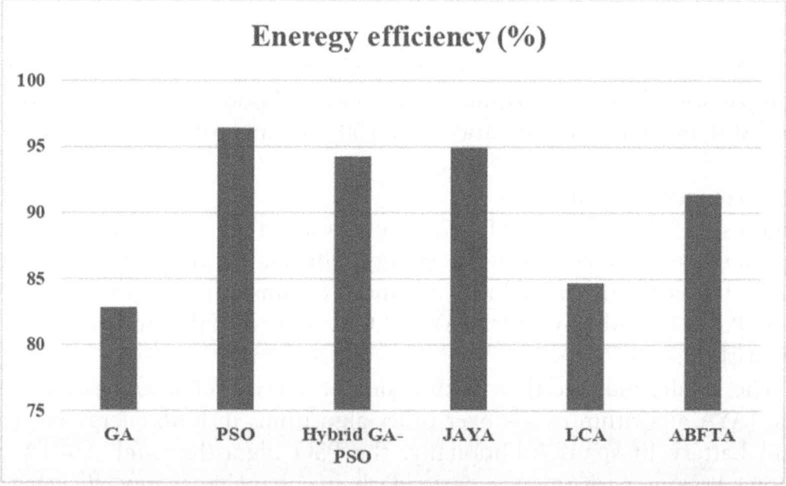

Figure 12.3 Comparison of optimization algorithm for energy efficiency parameter.

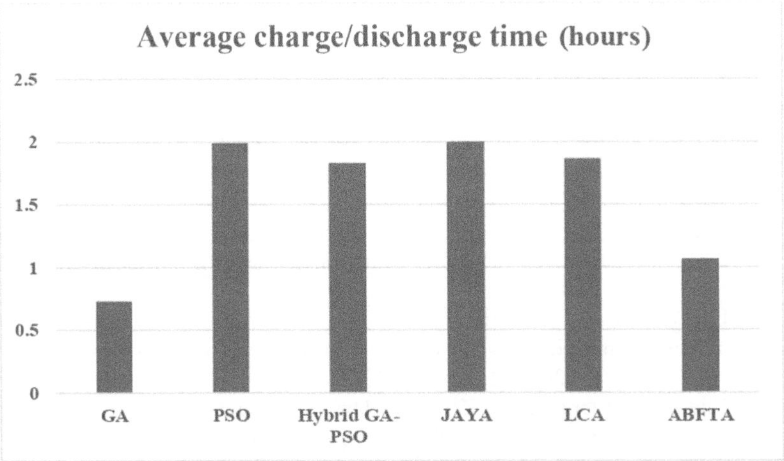

Figure 12.4 Comparison of optimization algorithm for average charge/discharge time parameter.

12.8 CONCLUSION

The study shows that integrating AI optimization techniques significantly enhances the performance of BMS in EVs. Among the algorithms analyzed, the JAYA algorithm stood out, offering the best balance between battery lifespan, energy efficiency, and charge/discharge times. These findings underscore AI's potential to optimize smart energy systems, paving the way for more sustainable and efficient EV technology. Future research could explore hybrid models that combine the strengths of various methods for even better optimization. Additionally, incorporating real-time data and machine learning could further enhance the performance and adaptability of BMS in EVs.

REFERENCES

1. He, H., Zhang, X., Wei, X., & Fan, J. Deep learning-based state-of-charge estimation for lithium-ion batteries. *Journal of Power Sources*, vol. 400, pp. 242–255, May 2018.
2. Berecibar, M., Gandiaga, I., Villarreal, I., Omar, N., & Van Mierlo, J. Machine learning techniques used to accurately predict battery life. *Journal of Power Sources*, vol. 320, pp. 267–278, July 2016.

3. Singh, N., Sharma, A. K., Sarkar, I., Prabhu, S., & Chadaga, K. IoT-based greenhouse technologies for enhanced crop production: A comprehensive study of monitoring, control, and communication techniques. *Systems Science & Control Engineering*, vol. 12, no. 1, 2024. https://doi.org/10.1080/21642583.2024.2306825
4. Singh, N., Sasirekha, S. P ., Dhakne, A., Sai Thrinath, B. V., Ramya, D., & Thiagarajan, R. IoT enabled hybrid model with learning ability for E-health care systems. *Measurement: Sensors*, vol. 24, pp.1–14, December 2022. https://doi.org/10.1016/j.measen.2022.100567
5. Zhang, L., Wang, J., & Li, X. Optimization of energy harvesting from renewable sources in electric vehicles using particle swarm optimization. *Applied Energy*, vol. 268, pp. 114–139, June 2020.
6. Liu, Z., Wu, J., & Zhang, W. Reinforcement learning for real-time energy management in electric vehicles. *IEEE Transactions on Vehicular Technology*, vol. 70(1), pp. 620–632, January 2021.
7. Chen, J., Huang, Y., & Zhang, J. Dynamic route optimization for electric vehicles using neural networks. *Transportation Research Part C: Emerging Technologies*, vol. 115, pp. 1026–1034, October 2020.
8. Sondhiya, V., Dwivedi, K., Kulsum Almas, S., & Singh, N. Energy management of nanogrid through flair of deep learning from IoT environments. In: *Artificial Intelligence Techniques in Power Systems Operations and Analysis*, Taylor & Francis, July 2023. ISBN: 9781003301820
9. Veena, R. S., Khambra, G., Sayal, R., Singh, N., Bala Dhandayuthapani, V., & Kumar, S. An energy efficient IPv6 packet delivery outline for industrial IoT over G.9959 protocol based wireless sensor network (WSN). *SN Computer Science*, vol. 5, p. 813, 2024. https://doi.org/10.1007/s42979-024-03195-4
10. Singh, N., Tamrakar, S., Mewada, A., & Gupta, S. K. *Artificial Intelligence Techniques in Power Systems Operations and Analysis* (1st ed.). Auerbach Publications, pp. 1–72, 2023. https://doi.org/10.1201/9781003301820
11. Liu, Y., Guo, J., & Li, H. Predictive maintenance for electric vehicle batteries using support vector machines. *IEEE Access*, vol. 7, pp. 48323–48329, April 2019.
12. Zhang, Q., Zhao, W., & Li, F. Early detection of motor faults in electric vehicles using random forests. *Mechanical Systems and Signal Processing*, vol. 104, pp. 665–678, February 2018.
13. Li, X., Wang, K., & Zhang, Z. AI and IoT for smart electric vehicles: A comprehensive review. *Journal of Cleaner Production*, vol. 265, 121802, August 2020.
14. Singh, S., Sharma, V., & Rana, R. Big data analytics and AI in smart electric vehicle systems. *IEEE Internet of Things Journal*, vol. 8, no. 2, pp. 802–810, February 2021.
15. Rohini, G., Gnana Kousalya, C., Singh, N., & Patil, V. R. Autonomous forecasting of traffic in cellular networks based on long-short term memory recurrent neural network. *Cybernetics and Systems*, pp.1–13, 2023.
16. Sondhiya, V., Dwivedi, K., Almas, S. K., & Singh, N. Energy management of nanogrid through flair of deep learning from IoT environments. In: *Artificial*

Intelligence Techniques in Power Systems Operations and Analysis, Taylor & Francis, July 2023. ISBN: 9781003301820
17. Singh, N. Optimization techniques used for design of economic electrical power distribution. In: *Meta Heuristic Optimization Techniques*, Walter De Gruyter GmbH, Berlin, Germany, ebook ISBN: 9783110716214, ISBN: 9783110716177, pp. 79–92, 31 January 2022. https://doi.org/10.1515/9783110716214-005

Chapter 13

Voting ensemble-based machine learning models for early detection of power system faults

Arvind Mewada, Maninder Singh, Harsh Pratap Singh, and Nagendra Singh

13.1 INTRODUCTION

Ensuring power systems' consistent and uninterrupted functioning is crucial for sustaining modern societies, industries, and economies. Power infrastructures are susceptible to various faults, including equipment malfunctions and unforeseen disruptions. These faults can lead to severe consequences, such as blackouts, equipment damage, and significant economic losses, highlighting the pressing need for effective fault detection and diagnosis in power systems [1, 2]. Addressing these challenges necessitates the adoption of advanced technologies and methodologies that enable early detection and precise identification of faults to ensure a stable and continuous electricity supply [3]. Traditionally, fault detection and diagnosis in power systems have relied on reactive measures, where operators respond to issues only after disruptions occur. This approach often results in delayed responses, making it difficult to prevent damage and minimise downtime [4]. A shift towards a more proactive and predictive fault detection system is essential to overcome these limitations. A proactive approach utilises real-time data to identify potential issues before they escalate, allowing operators to take timely corrective actions and maintain system reliability [5]. An essential aspect of maintaining a reliable power supply is power system stability, which refers to the ability of an electric power system to restore operational equilibrium following a disturbance. Stability ensures that critical system variables, such as rotor angles and bus voltages, remain within permissible limits, preserving the integrity of the overall network [6]. Power system stability can be categorised into three primary types: rotor angle stability, frequency stability, and voltage stability. Rotor angle stability involves synchronous machines maintaining synchronism after disturbances, while frequency stability pertains to the system's capability to maintain a consistent frequency despite significant imbalances between generation and load. Voltage stability ensures stable voltages are upheld across all buses following disturbances from an initial operating state [7].

DOI: 10.1201/9781003516156-13

Electric motor failures in industrial systems further exemplify the importance of effective fault detection. Such failures often lead to unexpected downtime, decreased production, increased operational expenses, and reduced profits [8]. Induction motors (IM), prevalent in households, commercial establishments, and industrial settings, are particularly susceptible to structural or component faults. The need for robust condition monitoring systems for IM is underscored by the potential consequences of sudden motor failures, which can endanger human lives and lead to costly production halts [9].

In recent decades, various knowledge-based systems, including Artificial Neural Networks (ANN), Genetic Algorithms (GA), Fuzzy Logic (FL), Particle Swarm Organization (PSO) [10–15], and Support Vector Machines (SVMs), have gained prominence due to advancements in technology and the availability of extensive datasets. These artificial intelligence (AI)-based methodologies provide rapid fault detection capabilities during real-time operations, minimising human error and enhancing situational awareness. Integrating smart meters and cost-effective sensors holds promise for further improving monitoring capabilities, allowing for precise fault location detection and management [12].

By adopting a predictive fault detection framework that leverages cutting-edge technologies such as advanced sensor networks and machine learning (ML) algorithms, this research addresses the pressing need for intelligent, data-driven monitoring solutions in power systems [11]. This proactive approach aims to empower power operators with timely and actionable insights, ultimately contributing to enhanced reliability and operational efficiency.

13.2 LITERATURE SURVEY

A noticeable variation in the frequency of voltage and current signals along a power transmission line occurs due to faults, and substantial shifts in this frequency can significantly impact interconnected power systems [13]. Detecting and analysing these changes promptly can prevent negative consequences, such as outages and system instability. Several methods have been utilised to analyse the time-domain attributes of frequency. Previous studies have employed the Fourier Transform (FT) for fault analysis, a widely utilised technique for both frequency and time-domain applications, which has been effective for fault classification [14, 15].

Additionally, the Half Cycle Discrete Fourier Transform (HCDFT) has been used for fault detection, while the Wavelet Transform (WT) has served as a robust tool for feature extraction [5]. Fault diagnosis in transmission lines has been conducted using the Discrete Wavelet Transform (DWT) and Continuous Wavelet Transform (CWT) [10]. However, these approaches often require considerable computation time to extract features from real voltage and current signals, increasing the computational burden on monitoring devices. To address this challenge, reference[16] developed a model for real-time fault detection that relies on a wavelet-based method, enabling fault detection without dependence on the selection of the wavelet. This approach calculates energy using wavelet coefficients obtained through a sliding window, contributing to reduced detection time.

Recent advancements in AI promise to enhance the accuracy and efficiency of fault analysis in power systems [17–20]. AI-based ML models, such as SVMs, K-nearest neighbours (KNNs), and Decision Trees (DTs), have become increasingly popular for transmission line fault detection [21, 22]. For instance, reference [23] achieved a fault detection accuracy of 99.42% using ML models, including SVM, KNN, and DT, and integrated DWT with SVM for fault classification, achieving an accuracy of 99.4% and a fault detection error of less than 1.5% [24]. Other studies, such as [25–27], have also successfully employed SVM for fault detection, utilising FFT for feature extraction.

An ML model was constructed combining a four-binary class SVM with WTs for fault identification in transmission lines [28]. ANN with DFT applied to phase current and voltage signals to identify and classify faults in power systems [29]. On the other hand, [30] combined DWT, DFT, and Particle Swarm Optimization (PSO) with ML models to effectively locate faults in transmission lines, using current signals of the zero set with KNN for fault detection.

Furthermore, reference [19] developed a model employing Bayesian linear discrimination as the classifier, utilising adaptive WT features for fault detection. Authors of reference [20] conducted a study on Power Quality Disturbance Classification utilising DWT and Multilayer Perceptron (MLP) to detect faults in microgrids. Reference [26] employed a DT with Self-Organized Maps (SOM) for fault location in transmission lines. Reference [22] explored supervised ML techniques requiring labelled data for classification. This two-stage process begins with data clustering using an unsupervised learning method, followed by a supervised learning technique for training and classification [23]. However, a significant drawback of this approach is its reliance on the outcomes of the unsupervised learning method, which can be heavily influenced by the selection of classification parameters.

13.3 PROBLEM FORMULATION

The dependable and uninterrupted supply of electricity is fundamental to the functioning of modern societies and economies. However, power systems are increasingly vulnerable to various faults and disturbances, jeopardising electrical grid stability and reliability. Unanticipated faults, which may arise from equipment failures or environmental factors, pose significant challenges to power system operators, potentially leading to service disruptions, equipment damage, and substantial economic losses. The current state of power system monitoring predominantly relies on reactive measures, where operators respond to faults only after they have manifested. This reactive approach often results in extended downtime and service interruptions. One of the critical issues facing power systems today is the need for more effective and proactive methods for early fault detection and diagnosis. Traditional monitoring systems frequently struggle to identify subtle precursors or patterns indicative of potential faults, leading to delayed responses and heightened vulnerability to system failures. The demand for advanced and proactive fault detection systems becomes paramount as power systems evolve, particularly integrating renewable energy sources, smart grids, and increasingly interconnected networks. The overarching problem addressed in this research is the need for a comprehensive and intelligent approach to early detection and precise diagnosis of faults in power systems. Existing methodologies often need to deliver timely information to operators, hindering their ability to take preventive measures and minimise the impact of faults. This research aims to develop a predictive fault detection framework that utilises advanced sensor networks and ML algorithms to identify potential issues before they escalate, enhance the accuracy and efficiency of fault diagnosis in power systems to enable proactive responses, and evaluate the effectiveness of the proposed methodologies through comprehensive experimental analysis and real-world case studies.

13.4 PROPOSED MODEL

The proposed methodology aims to improve power system fault detection and classification using an ML-based majority voting ensemble approach. This approach combines the predictions of various ML methods, including MLP, DT, KNN, SVM, Linear Regression (LiReg), Naive Bayes (NB), and Logistic Regression (LoReg). Each classifier was trained and tested using two separate datasets: one specifically for fault detection (the MLFD model) and the other for fault classification (the MLFC model). After training, each classifier produced accuracy scores and mean square error (MSE) values. The proposed model was developed, as depicted in Figure 13.1, to streamline the selection process, even for users with limited knowledge. This research presents multiple user options by integrating various ML approaches and offering recommendations based on different objectives. Users can choose classifiers based on high precision, recall, or F1 score criteria, facilitated by the ensemble method's average precision, recall, and F1 score metrics.

The research utilises two datasets from reference [17], each serving a specific purpose. The first dataset, presented in Table 13.1, focuses on detecting faults within the transmission line of a power system.

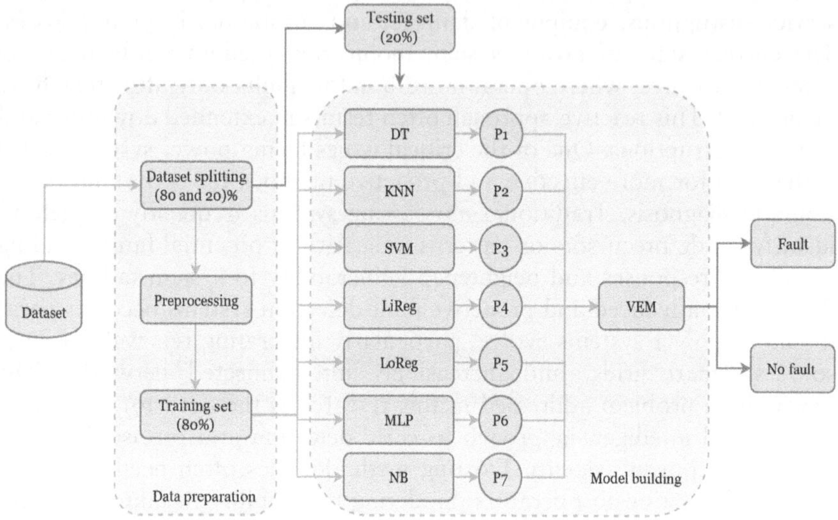

Figure 13.1 Proposed Voting Ensemble Model (VEM).

This dataset aids in identifying the presence or absence of faults, with the output being binary: '1' indicates the presence of a fault, while '0' signifies its absence. The Machine Learning-based Fault Detection (MLFD) model was trained using this fault detection dataset, as presented in Table 13.1.

The study utilises an additional dataset, as shown in Table 13.2, specifically focusing on fault classification within the power system's transmission line. This dataset identifies the presence of faults and categorises their locations within the transmission lines. For instance, the output 'GABC' indicates a line-to-ground fault between phases G and B, the output [0000] represents no fault, and [0010] denotes a fault in line B alone. The Machine Learning-based Fault Classification (MLFC) model was trained on this classification-focused dataset. After obtaining the dataset, it was divided into training and testing sets to train and evaluate different ML classifiers.

The datasets were divided into 70% for training and 30% for testing to ensure robust model evaluation. Subsequently, ML models were trained using seven different classifiers: KNN, Naive Bayes, SVM, LiReg, LoReg, DT, and MLP, integrated with a majority voting ensemble classifier. We developed two models, the MLFD model for fault detection and the MLFC model for fault classification, to perform training and testing. Each classifier individually calculated accuracy scores, and the MSE was computed and recorded for further analysis. These findings are presented in Tables 13.3, 13.4, and 13.5, thus providing valuable insights into the performance of each classifier for fault detection and classification3

Table 13.1 Sample dataset used in fault detection

S. no.	*Output (S)*	*Ia*	*Ib*	*Ic*	*Va*	*Vb*	*Vc*
0	0	−170.472196	9.219613	161.252583	0.054490	−0.659921	0.605431
1	0	−122.235754	6.168667	116.067087	0.102410	−0.628612	0.526202
2	0	−90.161474	3.813632	86.347841	0.141026	−0.605277	0.464251
3	0	−79.904916	2.398803	77.506112	0.156272	−0.602235	0.445963
4	0	−63.885255	0.590667	63.294587	0.180451	−0.591501	0.411050
...	...	...	...	...	...	...	...
11996	0	−66.237921	38.457041	24.912239	0.094421	−0.552019	0.457598
11997	0	−65.849493	37.465454	25.515675	0.103778	−0.555186	0.451407
11998	0	−65.446698	36.472055	26.106554	0.113107	−0.558211	0.445104
11999	0	−65.029633	35.477088	26.684731	0.122404	−0.561094	0.438690
12000	0	−64.598401	34.480799	27.250065	0.131669	−0.563835	0.438500

Note: Ia = Current in line A, Ib = Current in line B, Ic = Current in line C, Va = Voltage in line A, Vb = Voltage in line B, Vc = Voltage in line C.

Table 13.2 Sample of dataset used in fault classification

S. no.	*G*	*C*	*B*	*A*	*Ia*	*Ib*	*Ic*	*Va*	*Vb*	*Vc*
0	1	0	0	1	−151.29181	−9.677452	85.800162	0.40075	−0.13293	−0.26781
1	1	0	0	1	−336.1861	−76.283262	18.328897	0.31273	−0.12363	−0.18909
2	1	0	0	1	−502.8915	−174.64802	−80.924663	0.26572	−0.11430	−0.15142
3		0	0	1	−593.9419	−217.7033	−124.89192	0.23551	−0.10494	−0.13057
4	1	0	0	1	−643.6636	−224.1594	−132.2828	0.20953	−0.09555	−0.11398
...	...	...	...	...	...	...	...	...	...	...
7856	0	0	0		−66.23792	38.457041	24.912239	0.09442	−0.55201	0.45759
7857	0	0	0	0	−65.84949	37.465454	25.515675	0.10377	−0.55518	0.45140
7858	0	0	0	0	−65.446698	36.472055	26.106554	0.113107	−0.558211	0.445104
7859	0	0	0	0	−65.02963	35.477088	26.684731	0.12240	−0.56109	0.438690
7860	0	0	0	0	−64.59840	34.480799	27.250065	0.13166	−0.56383	0.43216

Table 13.3 Performance of various classifiers for fault detection

S. no	*Classifier*	*Accuracy (%)*	*MSE*
1	DT	99.40	0.006
2	KNN	99.12	0.0067
3	LiReg	90.00	0.095
4	LoReg	85.00	0.15
5	MLP	98.99	0.0032
6	NB	88.70	0.025
7	SVM	95.00	0.010

Table 13.4 Performance of various classifiers for fault classification

S. no	*Classifier*	*Accuracy (%)*	*MSE*
1	DT	87.00	0.31
2	KNN	85.00	0.45
3	LiReg	70.00	2.70
4	LoReg	70.00	1.50
5	MLP	77.99	2.40
6	NB	80.00	2.03
7	SVM	82.00	1.80

Table 13.5 Performance metrics of various models with Voting Ensemble Model

S. no	*Models*	*Precision (%)*	*Recall (%)*	*F1 Score (%)*	*Accuracy (%)*
1	DT	87.00	85.00	86.00	87.00
2	KNN	85.00	83.00	84.00	85.00
3	LiReg	59.00	60.00	59.50	70.00
4	LoReg	60.00	60.00	60.00	70.00
5	MLP	78.00	75.00	76.50	77.99
6	NB	80.00	78.00	79.00	80.00
7	SVM	82.00	80.00	81.00	82.00
8	**VEM**	**88.00**	**86.00**	**87.00**	**94.00**

The ensemble major voting method is implemented after collecting the accuracy scores and MSEs from all seven approaches considering the models (MLFD and MLFC). This method utilises prediction scores from the preceding classifiers to calculate the average prediction across all models. Consequently, it produces the ensemble accuracy score and MSE value. Each ML-based model produces eight MSEs and eight accuracy scores corresponding to the MLFD and MLFC models. Furthermore, precision, recall, and F1 score metrics are generated for all eight classifiers to evaluate the ML-detection model's performance comprehensively. These metrics are essential for accurately assessing the model's ability to detect faults. The resulting data is presented in Table 13.5, offering valuable insights to aid users in selecting the classifier that best suits their specific requirements.

13.5 EXPERIMENT SETUP AND ANALYSIS

A three-phase power system was simulated to generate datasets representing normal and faulty conditions. The simulation model included four 11 × 103 V position generators to simulate faults at the line midpoint. During the simulations, line voltage and current data were recorded for normal and faulty scenarios and labelled as 0 or 1 to represent the absence or presence of faults, respectively. This process resulted in two distinct datasets: a fault detection dataset and a fault classification dataset.

Fault detection dataset: This dataset comprised approximately 12,000 data points. The input features include line current and voltage values for the three phases: Ia, Ib, Ic, Va, Vb, and Vc. The output labels indicate the faults' presence (1) or absence (0) without specifying their location.

Fault classification dataset: This dataset contains around 7,860 data points and provides more detailed information, categorising faults into different types based on their location within the transmission line. Each record in this dataset is labelled using a specific format [G C B A], where 1 indicates a fault in a particular phase, and 0 indicates normal operation. For example, [1110] represents a double line-to-ground fault at the G, C, and B phases.

Both datasets were divided into training and testing sets using a 70:30 split ratio to ensure robust model evaluation. Seven different ML classifiers were initially trained using the designated training sets. The classifiers used included DT, KKNN, LiReg, LoReg, MLP, NB, and SVM.

Each classifier was evaluated based on its MSE, fault detection, and classification accuracy scores. The results are summarised in the following tables. Table 13.3 presents the performance of various classifiers for fault detection, the DT achieved the highest accuracy of 99.40% with an MSE of 0.006. The KNN closely followed with an accuracy of 99.12% and an MSE of 0.0067. The MLP also performed exceptionally well, achieving an accuracy of 98.99% and a low MSE of 0.0032. In contrast, the LoReg and LiReg classifiers had lower accuracy scores of 85.00% and 90.00%, respectively, with higher MSE values of 0.15 and 0.095. Additionally, the SVM attained an accuracy of 95.00% with an MSE of 0.010, while the NB classifier achieved an accuracy of 88.70% and an MSE of 0.025.

The SVM followed with an accuracy of 82.00%. LiReg and LoReg performed comparably, each achieving an accuracy of 70.00%. However, their precision and recall values were lower than those of the DT and SVM, indicating a different performance profile. The results emphasise the varied effectiveness of each model in classifying faults, providing valuable insights for selecting the appropriate algorithm based on specific performance criteria.

Figures 13.2 and 13.3 display the bar graphs for accuracy and MSE values, highlighting the performance of each classifier for fault detection. The results indicate that the DT achieved the highest accuracy of 99.40% with an MSE of 0.006, while the KKNN closely followed with an accuracy of 99.12% and an MSE of 0.0067. The MLP also performed exceptionally well, achieving an accuracy of 98.99% and a low MSE of 0.0032. In contrast, the LoReg and LiReg classifiers had lower accuracy scores of 85.00% and 90.00%, respectively, and higher MSE values of 0.15 and 0.095. The SVM attained an accuracy of 95.00% with an MSE of 0.010, while the NB classifier achieved an accuracy of 88.70% and an MSE of 0.025. For fault classification, Table 13.5 illustrates the performance of classifiers where the DT achieved the highest accuracy of 87.00% with an MSE of 0.31. The SVM followed with an accuracy of 82.00% and an MSE of 1.80. LiReg and LoReg performed comparably, each achieving an accuracy of 70.00%, but their MSE values were higher at 2.70 and 1.50, respectively. An ensemble method was introduced to improve model performance by combining the predictions of the base classifiers. The Voting Ensemble Method (VEM) aggregates the outputs of individual classifiers to produce a final prediction. This approach leverages multiple models' strengths, improving robustness and overall accuracy. Table 13.5 presents the performance metrics of various

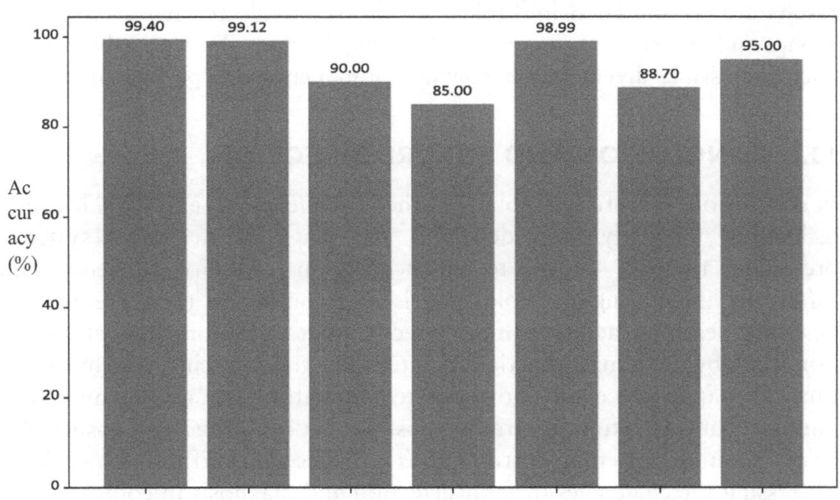

Figure 13.2 Accuracy of various classifiers for fault detection.

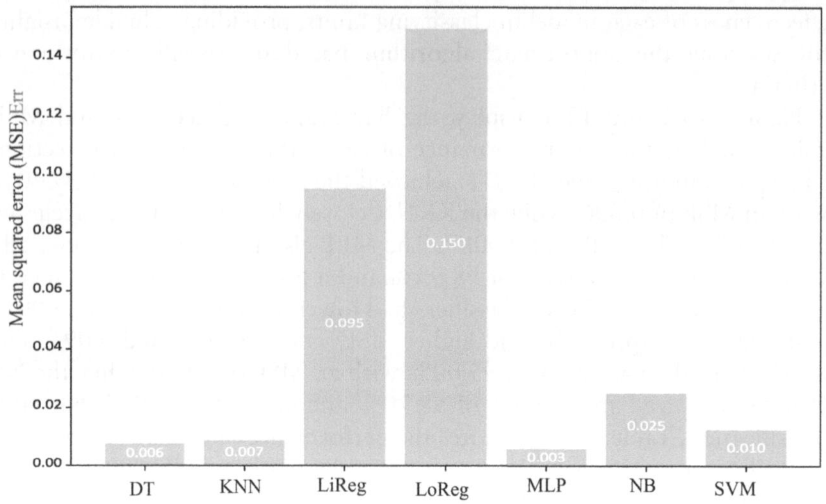

Figure 13.3 Mean square error of various classifiers for fault detection.

models with the VEM, outlining precision, reca3l, F1 Score, and accuracy values for each classifier, along with the ensemble method. The VEM achieved the highest performance with an accuracy of 94.00%, a precision of 88.00%, a recall of 86.00%, and an F1 Score of 87.00%. The individual contributions of each classifier are shown in Table 13.3, while the ensemble results are consolidated in Table 13.5. By randomly pairing classifiers in groups and observing their combined results, the study illustrates the importance of classifier diversity in achieving optimal ensemble performance.

13.6 CONCLUSION AND FUTURE DIRECTION

In conclusion, this study emphasises the effectiveness of ensemble learning techniques for early fault detection and diagnosis in power systems, presenting a robust strategy to enhance system reliability, efficiency, and safety. By integrating multiple models and combining their predictions, ensemble learning achieves more precise fault detection, and classification, thereby enabling proactive maintenance and reducing system downtime. This approach equips power system operators with diverse models to improve fault detection accuracy, boost system resilience, and ensure efficient operation. Ensemble learning thus provides a powerful framework for addressing the challenges of fault detection and diagnosis in complex electrical systems, ultimately contributing to grid reliability and stability.

Future research could incorporate advanced ensemble techniques such as gradient boosting and random forests and explore new ensemble architectures specifically tailored for power system fault detection. Developing adaptive algorithms that dynamically adjust to changing conditions would further enhance performance. Improving interpretability and explain ability by implementing multi-source data fusion methods, quantifying uncertainty, tackling scalability issues, and validating these systems in real-world environments are crucial directions for future exploration. Advancing these research areas can elevate the state of the art in early fault detection and ensemble-based diagnosis, leading to more reliable and resilient electrical infrastructure.

REFERENCES

1. O. M. Babatunde, J. L. Munda, and Y. J. Hamam. Power system flexibility: A review. *Energy Reports*, 6:101–106, 2020.
2. N. Singh, A. K. Sharma, M. Tiwari, M. Jasiński, Z. Leonowicz, S. Rusek, and R. Gono. Robust control of SEDCM by Fuzzy-PSO. *Electronics*, 12(2):335, 2023. https://doi.org/10.3390/electronics12020335
3. J. A. Casey, M. Fukurai, D. Herńandez, S. Balsari, and M. V. Kiang. Power outages and community health: A narrative review. *Current Environmental Health Reports*, 7:371–383, 2020.
4. T. Garcia-Calva, D. Morinigo-Sotelo, V. Fernandez-Cavero, and R. Romero-Troncoso. Early detection of faults in induction motors—a review. *Energies*, 15(21):7855, 2022.
5. W. Li, C. Wen, J. Chen, K. Wong, J. Teng, and C. Yuen. Location identification of power line outages using my measurements with bad data. *IEEE Transactions on Power Systems*, 31(5):3624–3635, 2016. https://doi.org/10.1109/TPWRS.2015.2495214
6. R. Saidur. A review on electrical motors energy use and energy savings. *Renewable and Sustainable Energy Reviews*, 14(3):877–898, 2010.
7. Y. Merizalde, L. Herńandez-Callejo, and O. Duque-Perez. State of the art and trends in the monitoring, detection and diagnosis of failures in electric induction motors. *Energies*, 10(7):1056, 2017.
8. G. Rohini, C. Gnana Kousalya, Nagendra Singh, and Visal Ratnsing Patil. Autonomous forecasting of traffic in cellular networks based on long-short term memory recurrent neural network. *Cybernetics and Systems*, 56(1):33–45, 2023. https://doi.org/10.1080/01969722.2023.2166245Berecibar, M., Gandiaga, I., Villarreal, I., Omar, N., & Van Mierlo, J. Machine learning techniques used to accurately predict battery life
9. Yunis Khan, A. Tevatia, D. Apparao, Nagendra Singh, and S. Mishra. Experimental investigation to evaluate thermal performance of a solar cooker with evaluated tube solar collector using different heat transfer fluids. *Journal of Thermal Engineering*, 10(5):1346–1357, September 2024. https://doi.org/10.14744/thermal.0000873

10. N. Singh, T. Chakrabarti, P. Chakrabarti, M. Margala, A. Gupta, S. B. Krishnan, and B. Unhelkar. A new PSO technique used for the optimization of multi objective economic emission dispatch. *Electronics*, 12(13):2960, 2023. https://doi.org/10.3390/electronics12132960
11. N. Singh, T. Chakrabarti, P. Chakrabarti, V. Panchenko, D. Budnikov, I. Yudaev, and V. Bolshev. Analysis of heuristic optimization technique solutions for combined heat-power economic load dispatch. *Applied Sciences*, 13(18):10380, 2023. https://doi.org/10.3390/app131810380
12. N. Singh, T. Chakrabarti, P. Chakrabarti, M. Margala, A. Gupta, S. P. Praveen, S. B. Krishnan, and B. Unhelkar. Novel heuristic optimization technique to solve economic load dispatch and economic emission load dispatch problems. *Electronics*, 12(13):2921, 2023. https://doi.org/10.3390/electronics12132921
13. Nagendra Singh, M. Praveen Kumar, and B. Santosh Kumar. Effect of valve loading on the thermal power economic load dispatch using new elephant herding optimization. *International Journal of Recent Technology and Engineering (IJRTE)*, 7(6S4):345–349, April 2019.
14. N. Torabi, V. M. Sundaram, and H. A. Toliyat. Online fault diagnosis of multi-phase drives using self-recurrent wavelet neural networks with adaptive learning rates. In *2017 IEEE Applied Power Electronics Conference and Exposition (APEC)*, pp. 570–577. IEEE, 2017.
15. Nagendra Singh. A new PSO techniques for the optimization of economic load dispatch of thermal generated power. *International Journal of Engineering & Technology*, 7(2):73–76, April 2018.
16. Anjali Nighoskar, Preeti Chaurasia, and Nagendra Singh. Advancing nuclear energy forecasting: Exploring regression modeling techniques for improved accuracy. *Nuclear Engineering and Technology*, 7(1):1–15, 2024. https://doi.org/10.1016/j.net.2024.08.013
17. N. Singh, S. Tamrakar, A. Mewada, and S. K. Gupta. *Artificial Intelligence Techniques in Power Systems Operations and Analysis* (1st ed.), pp. 1–72. Auerbach Publications, 2023. https://doi.org/10.1201/9781003301820
18. A. Jamehbozorg and S. M. Shahrtash. A decision-tree-based method for fault classification in single-circuit transmission lines. *IEEE Transactions on Power Delivery*, 25(4):2190–2196, 2010.
19. M. Rhif, A. Ben Abbes, I. R. Farah, B. Martínez, and Y. Sang. Wavelet transform application for/in non-stationary time-series analysis: A review. *Applied Sciences*, 9(7):1345, 2019.
20. S. Mohr, L. E. Ratcliff, P. Boulanger, L. Genovese, D. Caliste, T. Deutsch, and S. Goedecker. Daubechies wavelets for linear scaling density functional theory. *Journal of Chemical Physics*, 140(20), 2014. https://doi.org/10.1063/1.4871876
21. M. S. Chavan, N. Mastorakis, M. N. Chavan, and M. S. Gaikwad. Implementation of samlet wavelets to remove Gaussian additive noise from the speech signal. In Proceedings of Recent Researches in Communications, Automation, Signal Processing, Nanotechnology, Astronomy and Nuclear Physics: 10th WSEAS International Conference on Electronics, Hardware, Wireless and Optical Communications (EHAC'11), p. 37. Cambridge, 2011.

22. Nagendra Singh, Anjali Nighoskar, and Shiva Ram Krishna. Economic load dispatch with valve loading effects optimize by BBMO. *International Journal of Innovative Technology and Exploring Engineering (IJITEE)*, 8(7):104–107, May 2019.
23. A. Glowacz. DC motor fault analysis uses acoustic signals, coiflet wavelet transform, and k-nearest neighbour classifier. *Archives of Acoustics*, 40(3):321–327, 2015.
24. R. A. Sowah, N. A. Dzabeng, A. R. Ofoli, A. Acakpovi, K. M. Koumadi, J. Ocrah, and D. Martin. Design of power distribution network fault data collector for fault detection, location and classification using machine learning. In 2018 IEEE 7th International Conference on Adaptive Science and Technology (ICAST), pp.1–8. IEEE, 2018.
25. Nagendra Singh, S. P. Sasirekha, Amol Dhakne, B. V. Sai Thrinath, D. Ramya, and R. Thiagarajan. IOT enabled hybrid model with learning ability for E-Health Care Systems. *Measurement: Sensors*, 24:1–14, December 2022. https://doi.org/10.1016/j.measen.2022.100567
26. R. S. Veena, G. Khambra, R. Sayal, N.Singh, V. Bala Dhandayuthapani, and Sumit Kumar Sar. An energy efficient IPv6 packet delivery outline for industrial IoT over G.9959 protocol based Wireless Sensor Network (WSN). *SN Computer Science*, 5:813, 2024. https://doi.org/10.1007/s42979-024-03195-4
27. Nagendra Singh, Akhilesh Kumar Sharma, Indranil Sarkar, Srikanth Prabhu, and Krishnaraj Chadaga. IoT-based greenhouse technologies for enhanced crop production: A comprehensive study of monitoring, control, and communication techniques. *Systems Science & Control Engineering*, 12(1), 30 January 2024. https://doi.org/10.1080/21642583.2024.2306825
28. Nagendra Singh. Optimization techniques used for design of economic electrical power distribution. *Meta Heuristic Optimization Techniques*, pp.79–92. Walter De Gruyter GmbH, Berlin, Germany, 31 January 2022. https://doi.org/10.1515/9783110716214-005
29. K. Chen, C. Huang, and J. He. Fault detection, classification and location for transmission lines and distribution systems: a review on the methods. *High Voltage*, 1(1):25–33, 2016.
30. A. Glowacz. DC motor fault analysis uses acoustic signals, coiflet wavelet transform, and k-nearest neighbour classifier. *Archives of Acoustics*, 40(3):321–327, 2015.

Chapter 14

Solar PV array integrated UPQC for power quality enhancement

Siddharth Shukla, Amit Gupta, and Ruchi Pandey

14.1 INTRODUCTION

Power quality is a vital aspect of any electrical system, including microgrids. Microgrids, which are localized energy distribution systems, face unique challenges in maintaining a consistent and high-quality power supply. It is important to analyze the challenges faced in maintaining high-quality power supply within microgrids, which are crucial for ensuring reliable and efficient energy distribution. By exploring various aspects such as voltage fluctuations, harmonics, and frequency variations, this work sheds light on the key findings related to power quality issues in microgrids and also seeks to explore the significance of addressing power quality issues in microgrids to ensure their effective operation and contribution to the overall energy grid [1, 2].

The integration of solar photovoltaic (PV) systems with unified power quality conditioner (UPQC) technologies has gained significant attention in recent years due to the increase in clean and sustainable energy sources. This combination offers a promising solution to enhance power quality, mitigate voltage fluctuations, and improve overall grid performance [3]. By utilizing the renewable energy generated by solar PV systems and the power conditioning capabilities of UPQC devices, we can create a more reliable and stable power distribution network [4].

14.2 POWER QUALITY ISSUES IN MICROGRIDS

In today's world, the demand for cleaner and more sustainable energy sources is on rise. Microgrids have emerged as a promising solution to meet this demand by integrating diverse sources of energy[5]. However, with the implementation of microgrids comes the challenge of power quality issues. Let us delve deeper into the power quality issues that can arise in microgrids and explore potential solutions. The most common power quality issues in microgrids are discussed below [4–6].

 DOI: 10.1201/9781003516156-14

14.2.1 Voltage sags and swells

- Voltage sags and swells can occur due to sudden changes in the load within the microgrid system.
- These fluctuations can lead to equipment damage and disruption of operations.
- Implementing proper voltage regulation devices can help mitigate these issues.

14.2.2 Harmonic distortion

- Harmonic distortion is caused by nonlinear loads such as electronic devices.
- It can result in overheating of equipment and increased energy losses.
- Filtering solutions and active power filters can be employed to reduce harmonic distortion.

14.2.3 Frequency variations

- Frequency variations can impact the stability of the microgrid system.
- They can be caused by changes in load or sudden disconnection of a power source.
- Automatic controllers and energy storage systems play a crucial role in maintaining frequency stability [6].

14.2.4 Transients and voltage fluctuations

- Transients and voltage fluctuations can be caused by switching operations or lightning strikes.
- They can lead to equipment failure and data corruption.
- Surge protectors and voltage stabilizers are essential in safeguarding against these issues.

Power quality refers to the consistency and reliability of the voltage supplied to electrical equipment. Issues such as voltage sags, swells, harmonics, and interruptions can adversely affect the performance of sensitive devices. Poor power quality can result in equipment malfunction, data loss, and increased energy consumption [7].

14.3 ROLE OF UPQC IN POWER QUALITY OF POWER SYSTEM

UPQC is a modern power electronic device that plays a crucial role in improving power quality within electrical systems. By combining both series and shunt active power filters, UPQC can mitigate various power quality issues such as voltage sags, swells, harmonics, and flickers. The integration of UPQC in power systems not only ensures the efficient operation of electrical equipment but also enhances the overall stability and reliability of the system [8].

UPQC has the ability to improve power quality at the point of installation in power distribution systems or industrial power systems. Therefore, UPQC is expected to be one of the most powerful solutions for large capacity loads that are sensitive to supply voltage flicker/imbalance. With ideal compensation, the voltage at point of common coupling (PCC) is the fundamental positive sequence sinusoidal voltage on the power source side. The source currents are sinusoidal currents, and their phase angles are the same as the fundamental voltage in phase. In other words, with the UPQC function, the load is equal to a resistor [9]. As UPQC is a combination of series and shunt active filters, two active filters have different functions. The series' active filter suppresses and isolates voltage-based distortions. Active bypass filter cancels current-based distortions. At the same time, it compensates for the reactive current of the load and improves the power factor.

The significance of UPQC lies in its ability to provide a comprehensive solution to power quality problems by compensating for both voltage and current disturbances simultaneously. This advanced technology is increasingly being adopted in industrial, commercial, and residential sectors to ensure uninterrupted and high-quality power supply [9, 10]. Here, we analyze the applications of UPQC in different scenarios, including renewable energy systems, electric vehicle charging stations, and smart grids. By incorporating UPQC into these systems, we can effectively manage power fluctuations, improve system efficiency, and reduce energy losses.

14.3.1 Understanding UPQC

- UPQC is a power electronic device that mitigates various power quality issues, such as voltages, swells, harmonics, and power factor correction.
- It consists of two converters–series and shunt–that work together to regulate voltage and current at the PCC.

The UPQC system is divided into two parts. One is the power electronic converters module, which integrates the series and shunt converters, and the other is the series transformer, which handles the series injected voltage. Because of its superior performance in mitigating almost all major power quality problems, UPQC has become one of the most appealing solutions for improving power quality in the distribution system since then. Figure 14.1

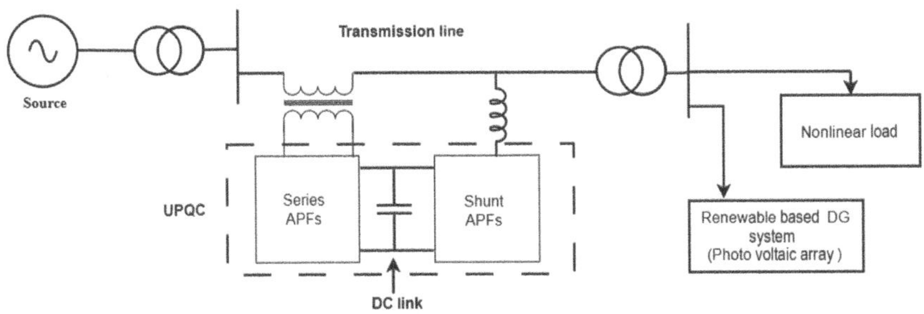

Figure 14.1 Basic structure of UPQC.

shows the structure and components of UPQC. However, UPQC is still in the research stage and has not been widely used thus far [11]. The reason for this is that the UPQC system includes two sets of power converters as well as a transformer. This significantly raises the costs of manufacturing and marketing the system. It is critical to lower the UPQC system's rating without compromising the compensating capabilities. In this way, it will increase UPQC utilization while lowering system costs, leading to improvements in the overall system's competitiveness [12].

14.3.2 Role of UPQC

- UPQC is a versatile device used to mitigate power quality issues.
- It combines both series and shunt active power filters to compensate for voltage distortions, sags, swells, and interruptions.
- UPQC not only protects sensitive loads but also helps in maintaining system stability and improving energy efficiency [12].

14.3.3 Benefits of UPQC integration

- *Improved power quality:* UPQC helps in maintaining voltage stability and reducing harmonic distortions, ensuring a reliable power supply.
- *Enhanced energy efficiency:* By correcting power factors and optimizing voltage levels, UPQC contributes to energy savings.
- *Increased system reliability:* UPQC minimizes voltage fluctuations and protects sensitive equipment from damage due to power disturbances.
- *Cost-effective solution:* While UPQC integration requires initial investment, the long-term benefits in terms of reduced downtime and maintenance costs outweigh the expenses.

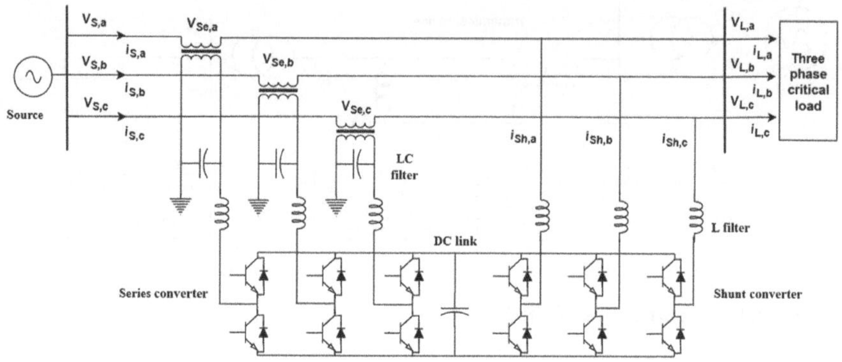

Figure 14.2 Schematic of UPQCs.

14.3.4 Future of power quality

- With the increasing integration of renewable energy sources and the electrification of transportation, the demand for high-quality power supply is on the rise [13].
- UPQC technology is continuously evolving to meet the challenges posed by modern power systems.
- Implementation of advanced control strategies and smart grid technologies will further enhance the effectiveness of UPQC in maintaining power quality.

The UPQC system which can improve power quality on both the source and load sides is shown in Figure 14.2.

14.4 THE UPQC CONTROL STRATEGY

The control strategy can be separated into shunt strategy, series control strategy, and DC capacitor control [14].

14.4.1 Shunt control strategy

The shunt active power filter helps to show both the current and the reactive power for the compensation. It works as a controlled current generator which compensates the load current to force the source currents drained from the network to be sinusoidal, balanced, and in phase with the positive-sequence system voltages [15].

14.4.2 Series control strategy

The series active power filter is another control strategy used for voltage compensation. To make the PCC voltage sinusoidal and balanced, it produces the compensation voltage that the pulse width modulation converter generates and inserts in series with the supply voltage.

14.4.3 The DC voltage regulator

The DC side voltage will fluctuate during the compensation process as UPQC makes up for losses in switches and active power, among other things. The output voltage of the series active filter will not equal the compensation value if the DC voltage differs from the rating value. There will be an error in the compensation. With the shunt active filter, it is the same. To maintain a constant voltage, a control signal is generated by the DC voltage regulator. It compels the network to supply more active current to the shunt active filter [14].

The analysis of the continuous voltage regulation at the storage capacity's limits revealed that filtering and speed control of this voltage need to be balanced. Because of this, the proportional integrator (PI), the regulator under study, is better adapted to guarantee both an ideal cost and filtering characteristic.

14.5 SOLAR PV INTEGRATED UPQC

Solar PV-integrated UPQC systems have the potential to address various power quality issues such as voltage sags, swells, harmonics, and power factor correction. The solar PV system generates clean energy from sunlight, which can be used to offset the grid's power demand and reduce electricity costs. The UPQC device, on the other hand, can control and regulate the voltage and current levels, correct power factors, and mitigate disturbances in the distribution system. By integrating these two technologies, we can achieve a more efficient, sustainable, and resilient power network [16].

14.5.1 Solar PV integration

- Solar PV systems generate clean and renewable energy, but their intermittent nature can impact power quality.
- By integrating UPQC with solar PV systems, the variability in power output can be managed effectively, ensuring stable and quality power supply to end-users.
- This integration enhances the overall performance and reliability of solar power plants, making them more attractive for both residential and commercial applications.

14.6 BASIC TOPOLOGY AND DESIGN

PV-UPQC's power circuit includes shunt and series compensators (essentially voltage source converters (VSCs)), also known as series active power filters (SAPFs) and parallel active power filters (PAPFs), which are linked by a DC link capacitor (which works as a DCbus). The shunt compensator is connected to the load side, whereas the series compensator is connected to the grid side via an injection transformer in each phase, which injects the voltage produced by the series compensator into the grid. The PV array is directly connected to the DCbus via a diode, which inhibits power transfer in the opposite way.

The series compensator, which is in voltage control mode, compensates for sags or swells in the grid side voltage. Interconnecting inductors are installed between the grid and both converters. The switching action of the converters might produce harmonics in the system, which are removed by the resistive-capacitive ripple filters. A bridge rectifier with a resistive-inductive load is linked to the load side and serves as a nonlinear load. PV-UPQC design includes PV module selection and sizing, maximum power point tracking algorithm selection, DC-link capacitor design, and DC-link voltage level selection. The next step is to develop interfacing inductors, injection transformers, and ripple filters [17]. Because the MPP voltage must be equal to the DC link voltage, the size of the PV module is determined by the DC link voltage. The shunt compensator should be sized so that it can handle the maximum power output of the PV array while also compensating for reactive power and load current harmonics. Another study [10] describes the design of parameters for a single-stage PV-UPQC. The ripple filters are chosen based on reference [12]. The most effective method for solving the power quality issue is the UPQC. The UPQC's overall setup is shown below.

When installed on industrial power systems or power distribution systems, the UPQC can enhance the quality of the power. Therefore, it is expected that the UPQC will be among the most effective ways to address big capacity loads that are susceptible to supply voltage fluctuation or imbalance. The voltage at PCC is the fundamental positive sequence sinusoidal voltage of the power supply side with perfect correction. The source's currents are sinusoidal, and each of their phase angles corresponds to the fundamental voltage in phase. Put another way, the load is equivalent to a resistance when the UPQC is in operation. As the UPQC combines shunt and series active filters, the roles of the two active filters are distinct. Voltage-based distortions are suppressed and isolated by the series active filter.

Current-based distortions are eliminated by the shunt active filter.

In addition, it raises the power factor and balances the load's reactive current. While there are other control strategies to establish the reference values for voltage and current, the most well-known ones are the instantaneous active and reactive power theory [16]. Put another way, the load is equivalent to a resistance when the UPQC is in operation. As the UPQC combines shunt and series active filters, the roles of the two active filters are distinct.

14.7 CONCLUSION

Solar PV integrated UPQC offers a good solution to power quality challenges in electrical systems. By combining the benefits of solar PV and UPQC technology, we can achieve a sustainable and reliable power supply that meets the demands of the modern world. It is time to embrace this innovative approach and pave the way for a greener and more efficient energy future. Keep in mind, the key to a brighter tomorrow lies in the integration of clean energy sources with advanced power quality solutions like solar PV integrated UPQC. Let us work together toward a more sustainable and reliable energy ecosystem.

Solar PV integrated UPQC is a cutting-edge technology that combines solar PV systems with UPQC to improve power quality and increase energy efficiency. This innovative solution helps to mitigate power fluctuations and voltage sags in the grid, ensuring a stable and reliable electricity supply. By integrating solar PV with UPQC, users can maximize energy generation from renewable sources while also enhancing the overall system performance. This integration not only benefits individual consumers but also contributes to a more sustainable and resilient energy infrastructure. Solar PV integrated UPQC holds great promise for transforming the way we harness and utilize clean energy for a greener and more sustainable future.

REFERENCES

1. Borkar, E., Singh, N. "Power Quality Enhancement by PV-UPQC for Non-Linear Load," In: *Artificial Intelligence Techniques in Power Systems Operations and Analysis*, Taylor & Francis, July 2023.
2. da Silva, S. A. O., Alves Negro, F. Single-phase to three-phase unified power quality conditioner applied in single-wire earth return electric power distribution grids, *IEEE Transactions on Power Electronics*, vol. 33, no. 5, 2018.
3. Jain, A., Singh, N. Boosting of harmonic distortion using fuzzy and PID controlled active power filter, *International Journal of Current Trends in Engineering & Technology*, vol. I, no. VII, pp.1–7, 2015.
4. Hussain, I., Agarwal, R. K., Singh, B. Three-phase single-stage grid tied solar PV ECS using PLL-less fast CTF control technique, *IET Power Electronics*, vol. 10, no. 2, pp. 178–188, 2017.
5. Kulsum Almas, S., Singh, N. Review on power quality improvement for grid connected wind energy generation system, *International Journal of Grid and Distributed Computing*, vol. 13, no. 1s, pp. 106–112, July 2020.

6. Wu, T.-F., Nien, H.-S., Shen, C.-L., Chen, T.-M. A single-phase inverter system for PV power injection and active power filtering with nonlinear inductor consideration, *IEEE Transactions on Industrial Applications*, vol. 41, no. 4, pp. 1075–1083, July 2005.
7. Javadi, A., Hamadi, A., Woodward, L., Al-Haddad, K. Experimental investigation on a hybrid series active power compensator to improve power quality of typical households, *IEEE Transactions on Industrial Electronics*, vol. 63, no. 8, pp. 4849–4859, August 2016.
8. Raghuvanshi, S., Singh, N. Comparative analysis of 36, 48, 60 pulse AC–DC controlled multipulse converter for harmonic mitigation, *International Journal of Advanced Research in Computer Engineering & Technology (IJARCET)*, vol. 3, no. 4, pp. 1159–1163, April 2014.
9. Raghuvanshi, S., Singh, N. Multipulse converter for analyzing total harmonics distortion and its mitigation, *International Journal of Engineering and Technical Research (IJETR)*, vol. 2, no. 7, pp. 49–52, July 2014. ISSN: 2321-0869.
10. Krishna, S. R., Kumar, R., Gaurav, A., Singh, N., Sharma, M., Almas, S. K. "Intelligent Adaptive Systems and Methods Thereof for Household Energy Control and Management," In: *2024 IEEE 3rd International Conference on Electrical Power and Energy Systems (ICEPES), Bhopal, India*, pp.1–3, 2024. doi: 10.1109/ICEPES60647.2024.10653541
11. Devassy, S., Singh, B. Modified p–q theory based control of solar PV integrated UPQC-S, *IEEE Transactions on Industry Applications*, vol. 53, no. 5, September/October 2017. doi:10.1109/TIA.2017.2714138
12. Khadem, S. K., Basu, M., Conlon, M. F. Intelligent islanding and seamless reconnection technique for microgrid with UPQC, *IEEE Journal of Emerging and Selected Topics Power in Electronics*, vol. 3, no. 2, pp. 483–492, June 2015.
13. Loh, P. C., Guerrero, J. M., Lee, T. L., Chandorkar, M. Advanced control architectures for intelligent microgrids; Part II: Power quality, energy storage, and AC/DC microgrids, *IEEE Transactions on Industrial Electronics*, vol. 60, no. 4, pp. 12631270, April 2013.
14. Singh, N., Priyadarshi, N., Kumar, N. "Meta Heuristic Techniques Used to Improve the Performance of Static Converters," In: *International Conference on Innovative Development in Virtual and Rapid Prototyping (IDVRP-2022). AIP Conference Proceedings* 2771, 020027, 2023. https://doi.org/10.1063/5.0152913
15. Singh, N., Almas, S. K., Tirole, R., Jain, A., Shukla, S., Kumar, Y. "Analysis of Optimum Cost and Size of the Hybrid Power Generation System Using Optimization Technique," In: *2023 IEEE 12th International Conference on Communication Systems and Network Technologies (CSNT), Bhopal, India*, pp. 284–291, 2023. doi:10.1109/CSNT57126.2023.10134720
16. Singh, N., Tamrakar, S., Mewada, A., Gupta, S. K. *Artificial Intelligence Techniques in Power Systems Operations and Analysis* (1st ed., pp.1–72), 2023, Auerbach Publications. https://doi.org/10.1201/9781003301820
17. Jain, A., Singh, N. Shunt active power filter controlled for mitigation of harmonics, *International Journal of Advanced Technology & Engineering Research (IJATER)*, vol. 5, no. 6, pp. 23–27, November 2015.

Chapter 15

Sustainability and renewable energy integration

Hemant Amhia, Ashish Choubey, Hari Prasad Bhupathi, and Shriram Pathak

15.1 INTRODUCTION

Many problems with energy sustainability and safety are plaguing the current system of energy management. If not addressed immediately, these problems have the potential to cause economic and political upheaval. Increasing attention is being paid to the development of sustainable alternative renewable energy sources (RESs) due to the depletion of fossil fuel deposits and the environmental impact of combustion. Solar photovoltaic (PV), concentrating solar thermal, geothermal, wind, wave, and biomass are some of the RESs that have seen significant market expansion [1]. A growing number of countries and businesses are committing to diversifying their energy sources by increasing the use of RESs as shown in Figure 15.1. Generation under the conventional system is dependent on customer demand, and distribution and consumption must be balanced for the grid to function efficiently.

15.1.1 Significance and motivation for study

Blackouts occur when energy demand outstrips supply, which destabilises the grid and reduces power reliability in certain locations. Electricity is wasted at an alarming rate when demand is lower than supply. Producing the appropriate amounts of electricity at the correct times is critical for effective grid operation and better economic benefits. Furthermore, the urgent need to promote more sustainable habits and technology is magnified by the fact that we are witnessing catastrophic consequences of climate change.

According to the International Energy Agency (IEA) [2], the energy sector is responsible for producing around 65% of greenhouse gases (GHGs), indicating a significant shift in the industry. Renewable energy bases are becoming more significant in our fight against climate change, even though the energy sector is responsible for 66% of global GHG emissions [3]. However, widespread implementation was hindered by the unpredictable patterns of solar and wind energy generation caused by weather impacts.

DOI: 10.1201/9781003516156-15

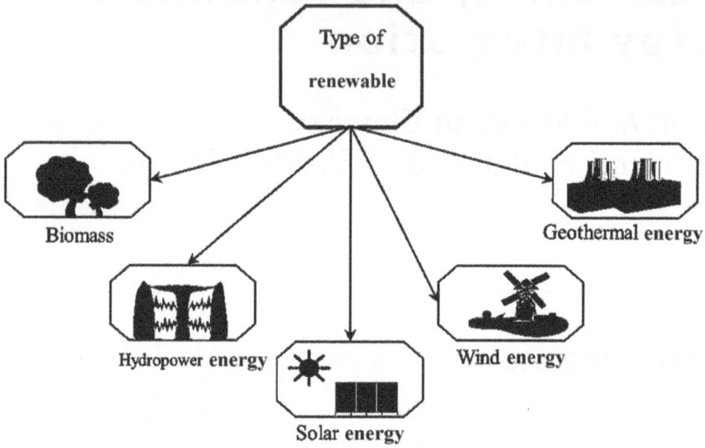

Figure 15.1 The renewable energy sources.

As a result, the electricity grid requires additional research and analysis to ensure a balance between supply and demand, prevent power outages, and provide enough energy to all users.

Energy generation and load prediction have been the subject of extensive study [4] to establish the amount of energy required to maintain stability, which in turn might lower operating costs and promote technology penetration with sufficient precision [5].

Improving short-term forecasting and incorporating this data into intelligent control systems that can optimise smart grid power distribution is one way to tackle this problem. It is common for artificial methods to surpass their corresponding methodological approaches in renewable energy optimisation [6–8]. To integrate renewable energy generation into current energy systems effectively, this chapter aims to compare and synthesise machine learning (ML) algorithms for accurate forecasting.

Using sensitivity analysis for the site in question, we provide the results of our modelling and analysis of a hybrid renewable energy system with maximum renewable penetration, taking into account three distinct model configurations: DG-PV-Wind-Battery, PV-Wind-Battery, and DG-PV-Battery. Under Hybrid Optimisation of Multiple Energy Resources (HOMER), the model is fine-tuned to minimise Levelised Cost of Energy (LCE) and Net Present Cost (NPC). We also take into account the reliability restriction of LPSP=0. To achieve sustainability, the suggested methodology seeks a configuration that maximises the use of renewable sources while simultaneously satisfying the specified system dependability criteria with the lowest possible values of NPC and LCE.

15.1.2 Renewable integration encounters

Electricity generation from a power grid might vary owing to the very changeable landscape of renewable energy causes including wind, solar heat, and solar light. This is because a number of environmental factors, such as solar irradiance, wind speed at various levels, cloud cover, and others, affect the overall production of these plants. Solar radiation, for example, is only available during the day, which is another major limitation of RESs. Thus, it is necessary to generate electricity, store some of it for later use, and use a portion of it all at once when resources are available [9].

Energy generation must be carefully managed due to the high expense involved in storing solar and wind energy. It is common practice to use conventional energy sources, such as gas plants, to make up for energy shortages when renewable resource capacity is insufficient. ML approaches are necessary to enhance planning for energy generation and utilisation due to the aforementioned concerns. ML approaches used in renewable energy-integrated grid systems vary according to problem criteria and characteristics. Creating both short-term and medium-term demand is essential for a renewable energy power grid to account for a significant portion of the energy supply [10].

For example, recognising critical variable quantities like the appropriate number of revolving investments and stowage requirements would help simplify the development of an energy strategy. Because power plant output is dependent on numerous uncontrollable environmental variables, it is necessary to predict both the generation of renewable energy from power plants and the power plant's output. Consequently, it is necessary to predict environmental variables surrounding the power plant, including sun irradiance, wind speed, and wind direction. The usage of ML techniques in the field of sustainable energy generation and integration tackles another critical issue: the ideal placement, dimensions, and specifications of renewable power plants [11].

The intelligent grid's overall operations and management also make use of ML approaches for problems like defect detection, regulation, and more. The objective was accomplished by presenting and applying ensembles of ML procedures such as Random Forest (RF) and Extreme Gradient Boost (XGB) to data from the hilly north-eastern Indian state of Tripura, with Agartala as its capital. This state has very little renewable energy integration; therefore, it could be a good place to start.

15.2 THE EXISTING WORK DONE

As energy accounts for a disproportionate share of the world's GDP, future sustainability depends on the planning and administration of energy systems. In addition, to reduce GHG emissions, these power grids should mainly use RES conversion technologies [12]. Both solar and wind power are the utmost widespread renewable energy choices due to their abundance and low environmental impact [13].

Photovoltaic systems and wind turbines are part of RES-based technologies and have a reputation for having a long lifespan and requiring little maintenance. The creation of a robust system to account for variations in the amount of energy that may be extracted from renewable sources is, however, made necessary by variations in weather conditions. At present, India is the fourth-largest economy in terms of purchasing power parity (PPP), and the fifth-largest energy consumer in the world.

In 2020, coal will provide half of India's energy needs, while oil and natural gas will provide 44%. Hydel power would account for 3.5% and nuclear power for 2.5%. Currently, natural gas accounts for 8.6% of our energy consumption, but by 2020, that number might rise to 14% [14]. Scientists are currently investigating potential solutions that could bring the level down to a sustainable level. From 53% in 2011 to 62% in 2035, according to a global energy demand assessment, the energy demand is expected to surge.

Various renewable technologies are well-suited to rural and outlying locations as well as developing nations, which is especially important given the current trend towards large-scale renewable energy projects. To supply the growing energy demand, optimal utilisation of properties obtainable about us is the only conceivable alternative [15].

According to a number of studies, renewable energy can help the world's poorest countries achieve unprecedented levels of wealth. Furthermore, non-RESs, such as fossil fuel power plants, often have losses of 40%–65% due to their steam cycles, whereas RESs, such as solar panels and wind turbines, are far more efficient, resulting in a substantial decrease in the need for non-RESs [16]. Sun, one of the few remaining biological power plants, irradiates an estimated 175,000 TW of energy.

Significant energy security and economic benefits could be achieved with the efficient and rapid expansion of RESs. The incineration of fossil energies is a major contributor to environmental pollution; this would improve the health of nature and help slowly the rate at which the planet is warming. Compared to other energy sources, RESs have significant potential spread out over a lot of land. Concerns about climate change and global warming, along with rising oil prices, oil scarcity, and government backing, are driving up the demand for renewable energy.

Researchers have shown that solar power has the potential to meet the majority of global electricity needs in the next half-century, cutting down on harmful emissions of GHGs [17]. To address the severe power restrictions and cut pollution, solar energy is an interesting addition for growing green energy capability. The Indian government had set a goal of producing 20,000 MW by the year 2020. In the first solar park of its kind in Asia, India aimed to use solar thermo-mechanical systems and solar photovoltaic

(SPV) technology to produce 72.4 GW of renewable energy, or around 16% of the total capacity, by the year 2022.

The applications of SPV technology range from small residential installations to massive power facilities with capacity of 1–100 MW. SPV systems have been around for a while, and many smaller-scale home installations have been using them. In recent years, researchers have used ML techniques to estimate PV systems' efficiencies [18]. A lot of the time, while predicting sun irradiance or wind speed, statisticians will engage Artificial Neural Networks (ANNs) or Support Vector Regression (SVR) because of their excellent estimating capabilities.

To predict solar irradiance, two main approaches are used: data-driven and model-based. When estimating solar irradiance using a model, it is important to take into version ecological influences counting temperature, humidity, wind speed and direction, and altitude. Reliability of the prediction effects is high, but models will restrict their practical application [19]. To establish a link between meteorological variables and PV power production, the alternative data-driven approach employing ANN depends solely on the accessibility of solar data.

Solar radiation, azimuth angles, and dry bulb temperatures were determined to be successful predictors of PV generation in ANN research. Multilayer perception feed forward ANN is used to construct ANN models that include forecasted climatic factors for the next day [20] to provide a 24-hour power forecast. An upgraded back propagation (BP) and ANN knowledge model was used to test a robust one-day SPV energy prediction under different meteorological conditions.

A hybridised support vector machine (SVM) has been used to make short-term PV grid forecasts. The RF algorithm is another popular method for making predictions, and it works well. Based on the results, RF is the best ML algorithm for predicting PV generation [21]. Nevertheless, ML models are notoriously unstable because their performance is highly dependent on learning.

15.3 THE PROPOSED METHOD

To accomplish the objective of the study and achieve sustainability, we suggest a forecast framework that integrates solar and wind generation, as well as electricity demand. The following variables were used as inputs: weather conditions (e.g., temperature, pressure, humidity, air density), solar irradiation (in KW), sun elevation (in degrees), wind speed (10 and 50 m), wind direction, and electrical load (to be anticipated).

The research methodology consists of two parts as shown in Figure 15.2. Data preparation and compilation are covered in the first section. The second section delves into the specific algorithms and methodologies employed by ML.

15.3.1 Data pre-processing

The primary focus here is on data collection and processing to build a data-driven model that can reliably predict electrical load, wind speed and direction, and solar output. What follows is a rundown of the methods utilised to process the data used in this study, as well as the data itself. The Ninja renewable project provided the data used in this chapter for each type of model training and validation, including weather conditions and the amount of wind energy at the specified location [9]. PV energy output estimates also take climatic factors like sun irradiation into account.

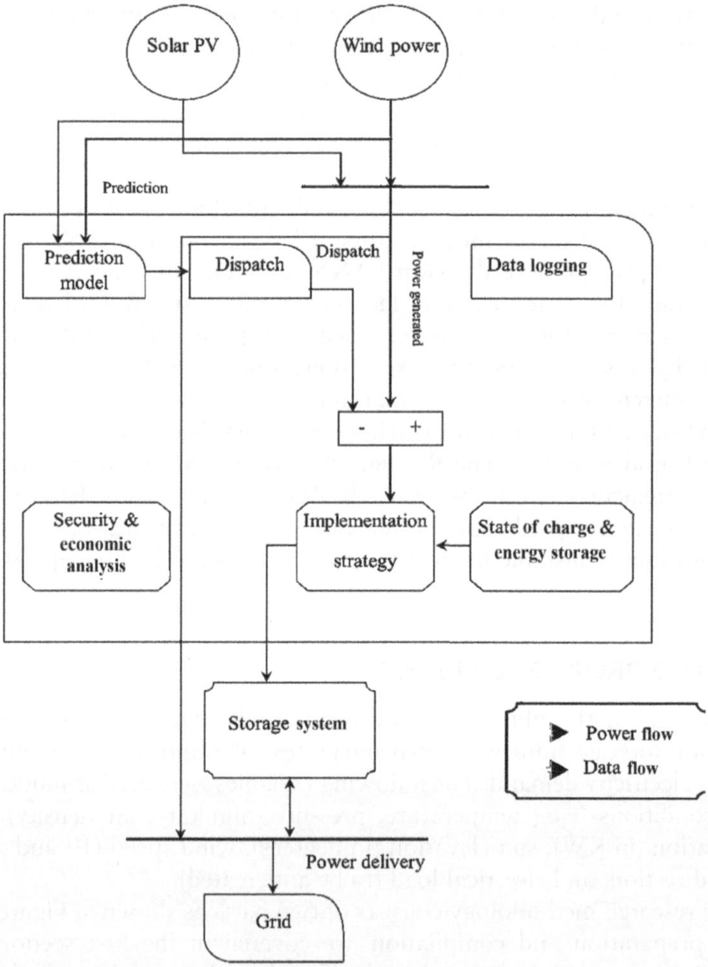

Figure 15.2 The proposed energy organisation system.

Lastly, the previous power load is the sole criteria utilised for electricity load projections.

15.3.2 Planning forecast

The input–output (IO) technique, the non-linear autoregressive (NAR) method, and the non-linear autoregressive with exogenous inputs (NARX) method are the three designs that can be used for data-driven time series estimation. The essential difference is in the kind of evidence that individually scheme takes into account as input. With the exception of the prior target series value, the IO will accept any and all inputs. When building the target, NARX takes together the previous values and external inputs into account, whereas NAR simply takes the prior values into consideration. As the NARX technique derives more knowledge about the system [12], it may be deduced to outstrip the additional two methods when the goal variable is connected to the exogenous inputs.

15.3.3 Ensemble model of RF and XGB in energy management system

The linked power system generates electricity by including RESs as an input. However, data from renewable sources can be detected and recorded across time series by the pyranometer and anemometer sensors. Additionally, the model predicts the amount of solar radiation and wind speed that will be present in the future using data from previous renewable sources. The section delves into the methods used by the RF and XGB ensemble to estimate the stochastic behaviour of electricity consumption, as well as wind and solar energy generation.

Using RF for feature selection helps to minimise complexity while addressing input selection concerns. We created a training set and a testing set from the dataset. We use the train set to choose features, which we subsequently apply to the test set.

Split energy: X-train, y-train, X-test, y-test = train-test-split (data, test-size = 0.3)

Feature-select = Select from Model (Random Forest Classifier (n-estimators = 100))

Feature-select.fit (X-train, y-train)

The anticipated outcomes are subsequently factored into the energy organisation scheme. A vital portion of any renewable power system's operation is the energy management system, which oversees all conditions for renewable energy generation as part of a vital power dispatching mechanism. Therefore, to project power and energy output, precise predictions of the future state of RESs are essential.

Its impact on the reliability of the power grid is substantial.

15.4 RESULT AND ANALYSIS

The methodology and procedures needed to forecast solar PV power are the same as those for wind energy forecasting. This is because sharing a preprocessing step is essential due to the shared source of the data. However, because SPV electricity is intermittent, the findings from this dataset are drastically different. To the right of each parameter is the Pearson correlation heat map that shows the relationships with solar energy.

SPV power output appears to be highly dependent on two factors: irradiance (radiation) and sun elevation [14]. While a characteristic "day_of_year" is generated and supplementary for upcoming training, a few factors are excluded from the heat map. Training is omitted for features with a much lesser impact on solar irradiance based on the projected heat map and feature selection.

Upon examination of the correlation heat map, it becomes apparent that there are correlations between a number of different factors. The two primary metrics utilised to estimate the presentation of the suggested model are the Root Mean Square Error (RMSE) and the (R^2) coefficient of determination. The R-squared value, which can take on values between 0 and 1, represents the percentage variation of the prediction results. Equipped with it is,

$$R^2 = 1 - \frac{\Sigma(\text{Actual} - \text{Predicted})^2}{\Sigma(\text{Actual} - \text{Mean actual})^2} \tag{15.1}$$

Reduced Mean Squared Error compares the model's actual values to its predictions. Finding its square root from the mean squared error yields the following formula:

$$\text{Reduced Mean Squared Error} = \sqrt{\Sigma \frac{(\text{Predicted} - \text{Actual})^2}{N}} \tag{15.2}$$

A collective comparative investigation of SPV power, wind energy, and electrical load prognostication is encapsulated in Table 15.1.

After conducting a comparative examination, it was determined that the (XGB)ensemble had greater values than other algorithms when it came to the forecasting of long-term values of wind energy, SPV power, and electrical load estimates. Consequently, for both short-term and long-term

Table 15.1 Comparative analysis of solar power, wind energy, and electrical load forecasting

		Wind energy		*Solar PV power*		*Electrical load*	
S. no.	*Prototypical*	*R^2-score*	*RMSE*	*R^2-score*	*RMSE*	*R^2-score*	*RMSE*
1	Boosted Decision Tree	0.623	1.312	0.835	5.426	0.382	3.456
2	Support Vector Regression	0.614	2.416	0.813	5.862	0.614	3.563
3	Random Forest	0.683	1.347	0.866	5.482	0.675	2.947
4	Boosted Random Forest	0.792	0.947	0.884	5.319	0.846	2.759
5	XG Boost ensemble	0.892	0.962	0.893	5.215	0.885	2.658

forecasting, the XGB ensemble has the potential to be utilised as a more suitable alternative for contemporary ensemble systems.

To address the matters connected to the generation and integration of RESs (solar and wind) through predictive modelling, a consistent dataset and a collaborative of RF for feature selection and XGB for training are utilised. An XGB and RFML ensemble is part of the model. To maximise power supply within a smart grid and achieve sustainability, a strategy is proposed to integrate the information in intelligent control systems and to improve the short-term forecast ability.

Concerning the generation and integration of RESs (solar and wind), a consistent dataset is utilised, along with an ensemble of RF for feature selection and XGB for training, to address the issues related to predictive modelling. Two ML ensembles of RF and XGB are part of the proposed model. This chapter lays out a plan to increase smart grid power supply efficiency and sustainability by integrating data into intelligent control systems and making short-term forecasts more accurate.

15.5 CONCLUSION AND FUTURE SCOPE

Renewable energy is becoming increasingly important for governments and corporations as outmoded energy sources like coal, oil, and natural gas become progressively scarce and the environmental consequences of using these fuels become more apparent. Exciting RESs like solar and wind have been expanding at a rapid pace. Due to the inherent unpredictability and lack of control over these resources' energy generation capabilities, proper management and forecasting methods are required to facilitate their integration into the grid. Integrating RESs into a system requires constant monitoring and the ability to foresee sudden shifts in energy demand and supply.

Here we see the state's electrical load, wind energy, and SPV power forecasted using predictive analytical methodologies. R2 and RMSE measure the dispersion of these forecasts. The provided optimisation model is used to regulate the finest architecture for a hybrid PV/wind system with battery storage based on the data values. The data-driven models are expected to improve power storage system investment and utilisation, which might lead to a financially feasible solution that allows for even higher rates of renewable power penetration into the grid.

Looking ahead to the energy market's potential, there are a number of new opportunities and challenges that must be addressed. The power industry may promote and contribute to the transition to new and sustainable technologies by incorporating all these possibilities into a prognostic smart grid regulator system and applying these investigate discoveries promptly. Further learning is desirable to test and comparability the limitations of the prevailing modelling. This research should include optimisation-based training approaches and more complicated models. Hence, this element necessitates further research.

REFERENCES

1. N. Singh, T. Chakrabarti, P. Chakrabarti,V. Panchenko, D. Budnikov, I. Yudaev, and V. Bolshev, "Analysis of heuristic optimization technique solutions for combined heat-power economic load dispatch," *Applied Sciences*, 13(18):10380, 2023. https://doi.org/10.3390/app131810380
2. R. Banik, P. Das, and S. Ray, "Prediction of electrical energy consumption based on machine learning technique," *Electrical Engineering*, 103(2):909–920, 2021.
3. K. A. Kavadias, and P. Triantafyllou, Hybrid renewable energy systems' optimisation. A review and extended comparison of the most-used software tools, *Energies*, 14(24), 2021. https://doi.org/10.3390/en14248268
4. A. Al Badi,and A. Al Wahaibi, "Techno-Economic Analysis and Optimization of Solar Energy System for Power Generation and Hydrogen Production in Al Mazyouna Area," In: *3rd International Conference on Clean and Green Energy Engineering (CGEE), Istanbul, Turkey*, pp. 26–31, 2022, IEEE. https://doi.org/10.1109/CGEE55282.2022.9976768
5. M. M. Rahman, G. A. Ghazi, E. A. Al-Ammar, and W. Ko, "Techno-Economic Analysis of Hybrid PV/Wind/Fuel-Cell System for EVCS," In: *International Conference on Electrical, Communication, and Computer Engineering (ICECCE), Kuala Lumpur, Malaysia*, pp. 1–6, 2021, IEEE. https://doi.org/10.1109/ICECCE52056.2021.9514084
6. S. Rehman, H. U. R. Habib, S.Wang, M. S. Büker, L. M. Alhems, and H. Z. Al Garni, Optimal design and model predictive control of standalone HRES: A real case study for residential demand side management, *IEEE Access*, 8:9767–9814, 2020, https://doi.org/10.1109/ACCESS.2020.2972302
7. C. A. Nallolla, and V. Perumal, "Optimal design of a hybrid off-grid renewable energy system using techno-economic and sensitivity analysis for a rural remote location," *Sustainability*, 14(22), article 15393, 2022, https://doi.org/10.3390/su142215393

8. B. K. Parida, and A. K. Bohre, "Optimal Sizing of PV, Wind and Biomass Based Grid Connected Hybrid Renewable Energy Systems for Rural Areas in Presence of EVs," In: *2nd International Conference on Emerging Frontiers in Electrical and Electronic Technologies (ICEFEET), Patna, India*, pp. 1–6, 2022, IEEE. https://doi.org/10.1109/ICEFEET51821.2022.9848154
9. S. K. Almas, and N. Singh,"Review on power quality improvement for grid connected wind energy generation system," *International Journal of Grid and Distributed Computing*, 13(1s):106–112, July 2020.
10. M. F. Ishraque, and M. M. Ali, "Optimized Design of a Hybrid Microgrid Using Renewable Resources Considering Different Dispatch Strategies," In: *International Conference on Automation, Control and Mechatronics for Industry 4.0 (ACMI), Rajshahi, Bangladesh*, pp.1–6, 2021. https://doi.org/10.1109/ACMI53878.2021.9528096
11. R. W. Moss, and R. Critoph, "Optimisation of a recirculating domestic hot water system to minimise wait time and heat loss," *Energy and Buildings*, 260, article ID 111850, 2022.
12. N. Singh, and Y. Kumar, "Multi objective Economic load dispatch problem solved by new PSO," *Advances in Electrical Engineering*, 2015:2356–6655, 2015. https://doi.org/10.1155/2015/536040
13. P. Valdiserri, "Evaluation and control of thermal losses and solar fraction in a hot water solar system," *International Journal of Low Carbon Technologies*, 13(3), 260–265, 2018.
14. S. K. Almas, and N. Singh, "Power quality improvement for grid connected wind energy. Generation system by implementation of DFIG scheme," *Turkish Online Journal of Qualitative Inquiry (TOJQI)*, 12(9):6928–6944, August 2021.
15. S. R. Krishna, R. Kumar, A. Gaurav, N. Singh, M. Sharma, and S. K. Almas, "Intelligent Adaptive Systems and Methods Thereof for Household Energy Control and Management," In: *2024 IEEE 3rd International Conference on Electrical Power and Energy Systems (ICEPES), Bhopal, India*, pp. 1–3, 2024. doi:10.1109/ICEPES60647.2024.10653541
16. N. Singh, A. Mishra, M. Gupta, K. Dwivedi, and M. Swarnkar, "Implementation of Microgrid for Rural Electrification to Cater on Side Load Demand," In: *2023 3rd International Conference on Energy, Power and Electrical Engineering (EPEE), Wuhan, China*, pp. 1058–1061, 2023, IEEE. doi:10.1109/EPEE59859.2023.10351810
17. N. Singh, S. K. Almas, R. Tirole, A. Jain, S. Shukla, and Y. Kumar, "Analysis of Optimum Cost and Size of the Hybrid Power Generation System Using Optimization Technique," In: *2023 IEEE 12th International Conference on Communication Systems and Network Technologies (CSNT), Bhopal, India*, pp. 284–291, 2023. doi:10.1109/CSNT57126.2023.10134720
18. N. Singh, S. Tamrakar, A. Mewada, and S. K. Gupta, *Artificial Intelligence Techniques in Power Systems Operations and Analysis* (1st ed., pp. 1–72), 2023, Auerbach Publications. https://doi.org/10.1201/9781003301820
19. S. F. Macedo, and D. Peyerl, "Prospects and economic feasibility analysis of wind and solar photovoltaic hybrid systems for hydrogen production and storage: a case study of the Brazilian electric power sector," *International*

Journal of Hydrogen Energy, 47(19):10460–10473, 2022. https://doi.org/10.1016/j.ijhydene.2022.01.133

20. A. El-Sattar, S. K. Hoda, H. M. Sultan, H. M. Zawbaa, and F. Jurado, “Optimal design of photovoltaic, biomass, fuel cell, hydrogen tank units and electrolyzer hybrid system for a remote area in Egypt,” *Energy Reports*, 8:9506–9527, 2022. https://doi.org/10.1016/j.egyr.2022.07.060
21. N. Singh, “Renewable energy for sustainable rural development and challenges,” *IJRAR*, 6(1):185–187, February 2019.

Chapter 16

Current and voltage asymmetry in 0.4 kV power grids and technical means to reduce them

Dmitry Pisarev and Alexey Bukreev

16.1 INTRODUCTION

The development of the industrial complex of different regions is directly related to the reliable and high-quality operation of 0.4 kV power distribution grids. For example, in different years, 0.4 kV power distribution grids transmitted from 8% to 15% of all electric power generated in Russia to consumers of various categories. This conditioned the approach to the construction of 0.4 kV power distribution grids based on the condition of minimum capital investments, often to the detriment of reliability of power supply and quality of electric energy. Modern industrial production requires high-quality power supply. Electricity quality requirements are not always met, although there are many regulatory documents establishing these requirements. Maintaining standardized power quality is becoming an increasingly urgent task, especially given the use of modern electrical equipment and automation systems that are sensitive to every indicator of power quality. Non-compliance with regulatory documents leads to economic damage, increased power losses, and other negative consequences. According to Borodin and Psarev [1], the problem of voltage asymmetry is traced to 0.4 kV grids and leads both to consumers' complaints about non-compliance of electricity supply with regulated standards and to large financial losses to electricity supply companies.

16.2 RESEARCH ON THE INFLUENCE OF CURRENT AND VOLTAGE ASYMMETRY ON THE QUALITY OF POWER SUPPLY AND LEVEL OF POWER LOSSES

Ensuring the quality of electric power for consumers and reducing power losses during its distribution are among the most difficult tasks that have to be solved in the design and operation of power supply systems. The degree of asymmetry of a three-phase four-wire system is characterized by asymmetry coefficients of currents and voltages in reverse and zero sequences. In this case, power losses due to asymmetry of currents can be characterized

DOI: 10.1201/9781003516156-16

by the power loss coefficient that is equal to the ratio of power losses in the asymmetrical mode to the corresponding power losses due to the flow of direct sequence currents. Voltage asymmetry coefficients and power losses in general can take any numerical values determined by the degree of voltage asymmetry. The larger they are, the greater the negative impact of the grid asymmetry on the operation of electrical installations, about which Bolshev, Papkov and others wrote in their studies [2]. According to Naumov, Shevchenko and Voyakin, power losses in 0.4 kV lines and 10/0.4 kV supply transformers with zero and reverse sequence current asymmetry coefficients equal to 25–30% increase by 30–50% compared to the symmetrical mode of grid operation [3].Voltage and current imbalances in electrical grids also lead to the following negative consequences:

- Increased power losses and high heating of neutral conductor
- Reduction of power supply reliability and occurrence of phase overloads
- Incorrect operation and increased wear of electrical machines
- Voltage drops in grids and, as a consequence, insufficient voltage for the operation of devices and instability of the grid as a whole
- Overvoltage is also a cause of failure of electrical equipment.

The block diagram presented in Figure 16.1 summarizes the negative consequences of the effects of stress asymmetry.

Elimination of asymmetry modes and minimization of power losses is one of the most significant problems, the solution of which requires the creation of new technical solutions to ensure the efficient operation of the industrial complex.

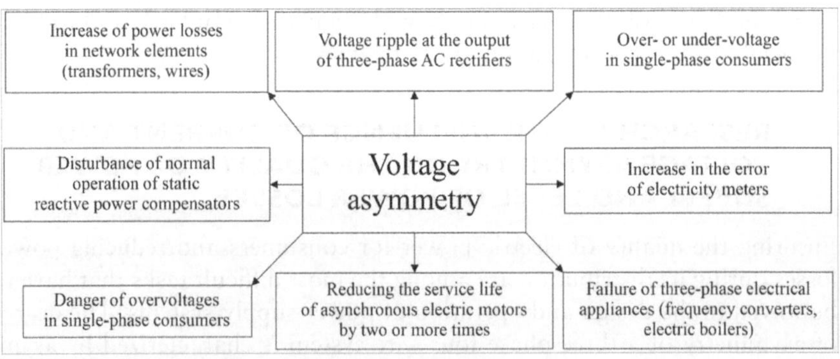

Figure 16.1 Negative effects of voltage asymmetry.

16.3 ANALYSIS OF CAUSES FOR CURRENT AND VOLTAGE ASYMMETRY IN 0.4 KV GRIDS

The mode of an electric distribution grid, when the parameters (current, voltage, power) of individual phases or an element of the grid are not equal, is called asymmetrical. Unsymmetrical modes were first described by Ch. Fortescue, who proposed a method for calculating these modes, which was described by Abdullaev and other authors [4].The symmetrical coordinate method is a mathematical method used to analyze and calculate three-phase electrical systems, especially in the context of asymmetrical conditions and unbalanced loads. It is based on the representation of three-phase alternating currents and voltages as their symmetrical components. The initial three-phase alternating currents and voltages are transformed into their symmetrical components: forward, reverse and zero sequence, presented in Figure 16.2.

I_A, I_B, I_C currents are decomposed into symmetrical components:

$$\begin{cases} I_A = I_1 + I_2 + I_3 \\ I_B = a^2 I_1 + a I_2 + a I_3, \\ I_C = a I_1 + a^2 I_2 + I_3 \end{cases} \tag{16.1}$$

where I_A, I_B, I_C: currents in phases; I_1, I_2, I_3: currents of symmetrical components; a:rotation operator $a = e^{j2\pi/3} = e^{-j4\pi 3} = -\frac{1}{2} + j\frac{\sqrt{3}}{2}$.

Solving the system of equations (16.1) with respect to I_1, I_2, I_3 we obtain:

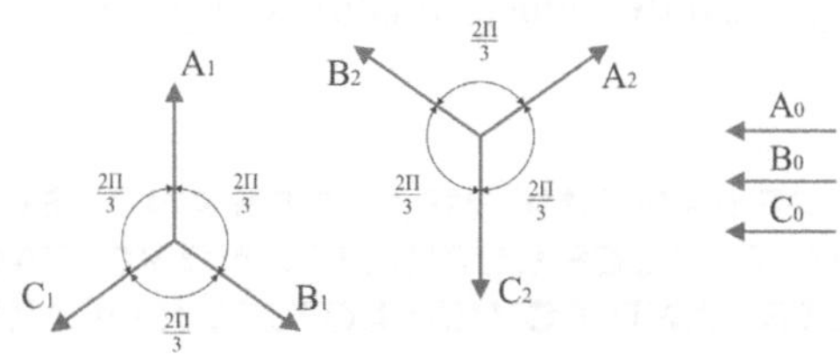

Figure 16.2 Vector diagrams of sequences: (a) forward; (b) inverse; (c) zero.

$$\begin{cases} U_A = U_1 + U_2 + U_3 \\ U_B = a^2U_1 + aU_2 + U_3 \\ U_C = aU_1 + a^2U_2 + U_3 \end{cases} \tag{16.2}$$

$$\begin{cases} U_1 = \frac{1}{3}\left(U_A + aU_B + a^2U_C\right) \\ U_2 = \frac{1}{3}\left(U_A + a^2U_B + aU_C\right), \\ U_0 = \frac{1}{3}\left(U_A + U_B + U_C\right) \end{cases} \tag{16.3}$$

where U_A, U_B, U_C: voltages in phases; U_1, U_2, U_3:voltages of symmetrical components. Unsymmetrical operation modes of power distribution grids can be divided into different types: short term and long term, accidental and systematic.

According to Zakurdaev and Chernetskaya, emergency processes in power supply systems, such as short circuits, and transmission line breaks, which lead to the occurrence of incomplete-phase modes of three-phase grids, are the causes of asymmetrical modes of short-term nature [5].

Prolonged asymmetrical modes appear in the case of phase-to-phase differences in system parameters resulting from the uneven distribution of single-phase receivers in three-phase grids and their probabilistic nature of switching. Until recently, incomplete-phase branch circuits were still a significant cause of long-term asymmetrical modes. At present, the influence of this cause is decreasing, as more and more three-phase connection of consumers to power grids is performed, according to Bolshev and Vinogradov [6]. Uneven distribution of single-phase electric consumers by phases leads to overloading of some phases and underloading of other phases, which results in systematic asymmetry. The cause of random (probabilistic) asymmetry is random switching on and off of individual single-phase electric consumers. In general, these causes lead to unbalanced consumption of electric energy, which significantly reduces quality and increases losses of electric energy.

16.4 REVIEW OF EXISTING METHODS AND TECHNICAL MEANS TO REDUCE CURRENT AND VOLTAGE ASYMMETRY AND POWER LOSSES IN 0.4 KV GRIDS

To reduce voltage asymmetry in 0.4 kV power grids, which have a deaf-earthed neutral, various methods and technical means have been developed. These methods differ in complexity, cost, and economy, as well as in efficiency

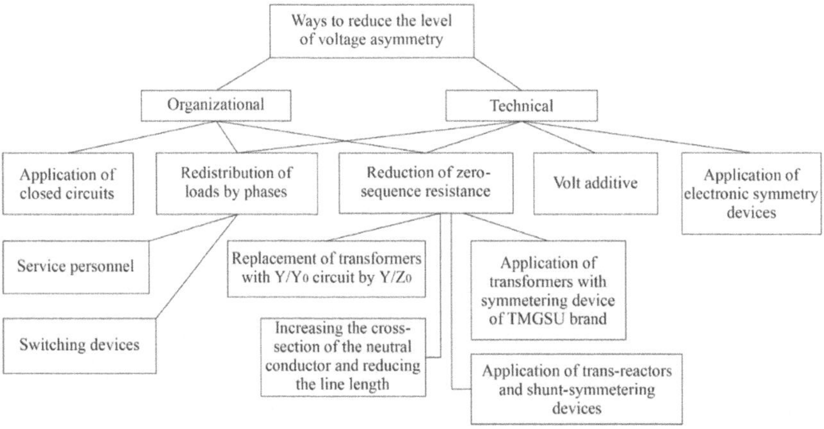

Figure 16.3 Classification of methods to reduce voltage asymmetry.

and principles of operation. These means are available for different categories of consumers. All these methods can be divided into two types: technical and organizational. The technical approach involves the development and introduction of certain devices into the existing electrical system and the reconstruction of the existing grid. The organizational approach implies carrying out various operational activities with the existing grid, using existing elements and devices, as well as the current structure of this grid. Certain decisions are made when designing new grids. Figure 16.3 shows the classification of ways to improve the quality of electric power according to the level of voltage asymmetry.

As shown in Figure 16.3, it is possible to reduce voltage asymmetry in 0.4 kV grids by different methods: application of closed circuits of 0.4 kV distribution grids, even distribution of load over grid phases, reduction of zero-sequence resistance of the grid, elimination of zero and reverse sequence components of three-phase grid voltage. Equalization of loads by phases can be performed manually or automatically with the help of special control devices.

Reduction of the zero-sequence resistance of the grid can be achieved by increasing the cross section of the neutral wire, reducing the length of lines, replacing the transformer with Y/Y_0 winding connection scheme at the 10/0.4 kV substation with a transformer with Y/Z_0 winding connection scheme or with a transformer with a symmetering device of TMGSU brand, or by introducing shunt-symmetering devices. Elimination of voltage asymmetry can be achieved by introducing various volt adders. Electronic symmetry devices, including inverter devices, also allow solving the problem of voltage asymmetry in 0.4 kV grids effectively.

16.4.1 Load redistribution by grid phases

When designing and introducing new consumers into 0.4 kV grid phases, they strive to distribute single-phase consumers evenly across the phases. However, the load generated by each of them depends on the time of day and season and increases from year to year in individual residential buildings, which leads to voltage asymmetry. To reduce asymmetry, the load can be allocated to the phases based on phase current measurements every 6 months. In addition to manual switching of single-phase consumers by service personnel, automatic devices can be used based on continuous monitoring of current ratios in different grid phases. However, the approach to reducing voltage asymmetry through periodic manual or automatic load sharing has a disadvantage: due to the variable nature of the load, the result will not be stable and will depend on the choice of methodology and the order of measurements.

16.4.2 Application of closed circuits of distribution grids

Scientific research has shown that the use of fully or partially closed systems in electrical grids can improve their operation. The load on individual grid phases becomes more balanced, and power and voltage losses and voltage asymmetry in the grid are reduced. When two lines are connected, the expected zero or reverse-sequence voltage is 33% lower than the maximum observed value in unconnected radial lines. If several power transformers with different secondary bus voltages are connected to the grid, equalizing currents can occur, leading to increased power losses, according to Churin and Sukyasov [7].

16.4.3 Reduction of zero-sequence resistance of the grid

In rural distribution grids with deaf-earthed neutral voltage of 0.4 kV, the value of zero-sequence resistance is 10–15 times higher than the value of direct sequence resistance. Therefore, the zero-sequence component of phase voltages will have a significant value even with fairly small values of zero-sequence current I_0, creating a large asymmetry in phase voltages, according to Naumov, Fedorinova and Yakupova [8]. Hence, symmetry of voltages at consumer inputs can be realized by reducing the value of the zero-sequence voltage U_0 by reducing the value of the zero-sequence resistance Z_0 of the grid. This is well illustrated by the following formula:

$$U_0 = I_0 \cdot Z_0 \tag{16.4}$$

The complex resistance of the zero sequence of the grid Z_0 is equal to the sum of similar resistances of the transmission line Z_{L0} and power supply transformer at the substation 10/0.4 kV Z_{T0}, which determines the possible ways to reduce the value of zero-sequence resistance of the grid, according to Wang, Gao and others [9]. The easiest way to achieve a reduction in the total resistance Z_0 is to reduce Z_{T0}.

In rural electrical grids at transformer substations (TSs) of 10/0.4 kV, power transformers with winding connection scheme Y/Y_0 ("star-zest with zero"), which have the highest value of zero-sequence resistance among existing transformers with other winding connection schemes, are currently widespread. In Y/Y_0 transformers, uncompensated magnetic fluxes from the zero-sequence currents penetrate the steel tank, causing its heating. Due to the law of electromagnetic induction, these fluxes produce an electromagnetic force (EMF) in the windings against the zero-sequence currents, which determines the increased zero-sequence resistance of such transformers.

The situation with the ZT_0 value is much better for power transformers with winding connection schemes Δ/Y_0 ("delta-star-zero"), Y/Z_0 ("star-zigzag-zero"), and especially for TMGSU transformers (power transformers with a symmetering device). In the case of Y/Z_0 transformers, the resistance ZT_0 is 9–11 times smaller than that of Y/Y_0 transformers. The simplified structure of the transformer with Y/Z_0 winding connection scheme is shown in Figure 16.4.

The magnetic conductor of such three-phase transformers is made of pressed plates of electrical steel and has three rods. The primary winding is connected in a typical star pattern without a neutral conductor. The secondary winding is connected in a zigzag pattern: each rod carries one-half of two different secondary windings. In this way, the magnetic fluxes in the asymmetrical mode are mutually compensated, which reduces the zero-sequence resistance, according to Wang, Gao and others [9.

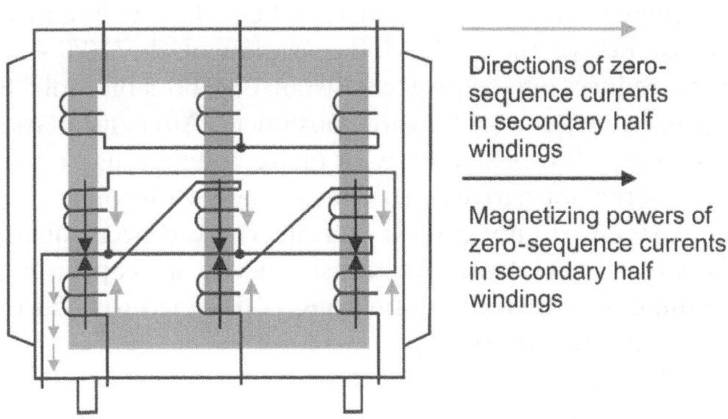

Figure 16.4 Y/Z_0 transformer structure and processes in the transformer.

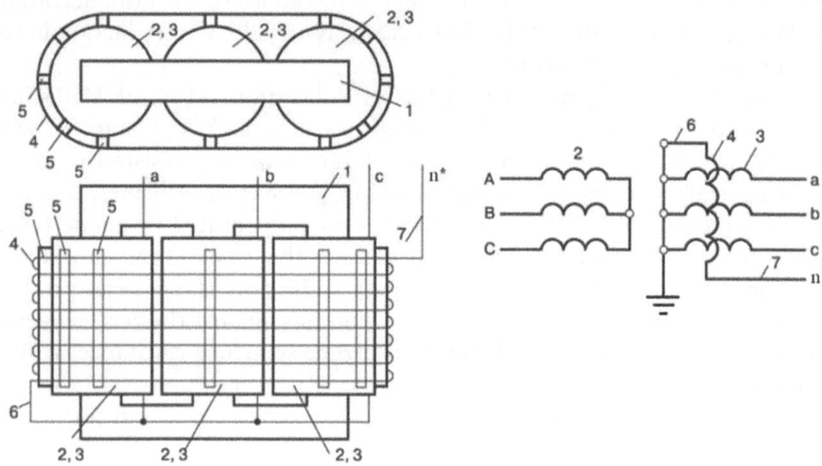

Figure 16.5 TMGSU transformer design and its schematic diagram: 1–magnetic conductor; 2 and 3–primary and secondary windings of the power transformer; 4–additional bandage winding; 5–wedges; 6–output from the zero point of the transformer windings; 7–output of additional bandage winding for connection to the neutral wire of the grid.

Transformer (three-phase) oil sealed and equipped with a symmetrical device (TMGSU transformer), as shown in Figure 16.5, is an advanced power transformer with a Y/Y_0 winding connection scheme.

In transformers with Δ/Y_0 winding connection, the load components of the primary winding currents from the secondary winding zero-sequence currents are counter-rotated, which reduces the value of the zero-sequence resistance of the grid Z_0. Therefore, it is possible to replace transformers featuring winding connection scheme Y/Y_0 with transformers with winding connection scheme Y/Z_0, Δ/Y_0 or with TMGSU, which will reduce the zero-sequence resistance of the grid. Modernization of the 10/0.4 kV supply substation to reduce zero-sequence resistance is possible only for power supply organizations, not for private consumers. After installation of new transformers, the voltage asymmetry of phase voltages may remain significant and exceed the normative level. Transformers with connection schemes Y/Z_0, Δ/Y_0, TMGSU do not protect consumers from over voltages in case of neutral conductor breakage. The transformer with a capacitor in the secondary winding, as shown in Figure 16.6, is designed to reduce the zero-sequence resistance of the grid Z_0.

To reduce the total zero-sequence impedance, the primary winding of a transformer with a secondary winding connected to a capacitor bank can be inserted into the neutral conductor break. In this arrangement, a current

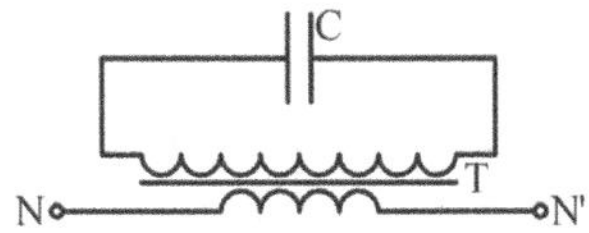

Figure 16.6 Transformer with capacitor on secondary winding.

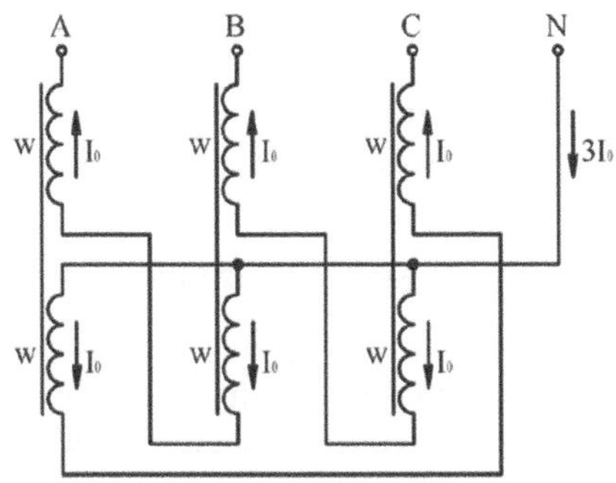

Figure 16.7 Electromagnetic shunt-symmetering device.

of active-capacitive character will flow through the primary winding of the transformer. This makes it possible to compensate for the reactive inductive component of the total zero-sequence resistance. Such transformers have two disadvantages. The first is that the active component of the zero-sequence resistance of the grid is not compensated when using long rural grids with utility load.

Second, in case of a break in the secondary winding circuit or damage to the capacitor bank, the primary winding introduces additional reactive resistance, which worsens the situation with the level of asymmetry, which is unacceptable according to the rules of technical operation of the grid. The resistance Z_0 of the grid can be reduced by connecting at the beginning or end of the transmission line a shunt-symmetering device, which has a small value of resistance to zero.

This device shunts the zero-sequence currents, preventing current from flowing into the line and the power transformer. This makes the zero-sequence voltage at the load as low as possible, determined by the resistance of the Shunt-Symmetering Device (SSD). The three-phase choke, as shown in Figure 16.7, is an electromagnetic shunt-symmetering device. Two windings with the same number of turns are placed on each rod of the electrotechnical steel magnetic core, which are included in the scheme "counter zigzag". For direct sequence, currents of the choke windings are characterized by a significant inductive resistance, and for zero-sequence currents–a small active resistance. In general, this provides a low zero-sequence resistance of the electromagnetic SSD.

Zero inductive resistance for zero-sequence currents is explained by the "counter zigzag" winding connection scheme, when zero-sequence currents create counter-directed magnetic fluxes, mutually subtracting and giving a total zero-sequence magnetic flux equal to zero. However, the electromagnetic SSD has the disadvantage that it reduces the power factor of the grid. The parameters of shunt-symmetering devices do not depend on the value of the load, which is their advantage compared to other symmetering devices. Both capacitor and inductive shunt symmetry devices significantly, but not completely, reduce voltage asymmetry. Moreover, their parameters depend on the reactive power of the load.

16.4.4 Three-phase voltage stabilizers

Three-phase voltage stabilizers are often used for automatic correction of voltage levels by phase, in cases of supplying three-phase power consumers, or when connecting a group of single-phase appliances to a three-phase grid phase. Currently, three-phase stabilizers are known, as shown in Figure 16.8. These are ferroresonant, electromechanical, electronic (relay and thyristor), with magnetized transformers, inverter, electric machine.

In electric machine stabilizers, an asynchronous motor connected to a three-phase power supply drives the shaft of a three-phase synchronous generator. A flywheel is rigidly connected to the shafts of these machines. There is a conversion of electrical energy into mechanical energy, which is stored by the flywheel. Voltage stabilization is ensured by the inertia of the flywheel. Short-term changes in the input voltage do not affect the flywheel rotation speed and the generator output voltage. Such stabilizers fully

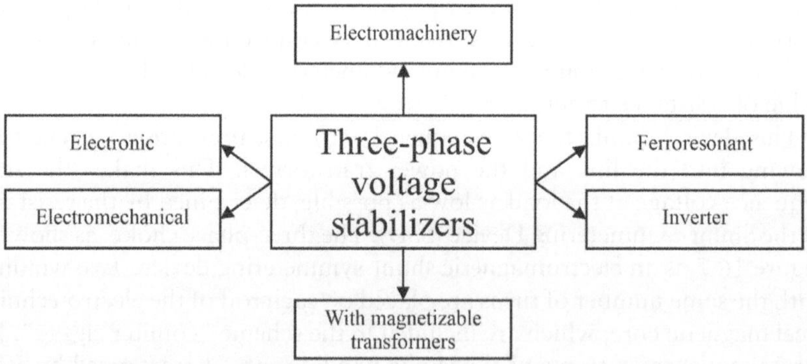

Figure 16.8 Types of three-phase voltage stabilizers.

restore the initial phases and the shape of phase voltages, but they have high material intensity, high noise level, high cost, and require qualified maintenance.

Electromechanical voltage stabilizers are devices based on toroidal magnetic conductors, with a part of the surface winding stripped of insulation. It provides electrical contact with a movable graphite roller, which acts as a current collector. The movable contact is moved over the surface of the winding by means of a servo drive, which is controlled by an electronic circuit. It controls the voltage level at the stabilizer input. Moving the movable contact changes the number of turns connected to the output while the number of turns to which the line voltage is applied remains unchanged. This changes the transformation ratio of the autotransformer and ensures that the stabilizer output voltage remains constant when the input voltage shows fluctuation.

However, compared to other voltage stabilizers, this type has a low regulation speed, which negatively affects the quality of the output voltage during rapid changes in line voltage. In addition, the sliding current collector of this type of stabilizers is subject to wear and requires special maintenance.

Electronic voltage stabilizers are usually based on autotransformers with many branches from the winding through a certain number of turns. The mains voltage is supplied to the part of the winding with a constant number of turns, and all branches are connected to the stabilizer output through electromagnetic relay contacts or semiconductor simistors. The main disadvantage of such stabilizers is discrete voltage regulation, which can cause noticeable changes in the brightness of light bulbs, leading to physical discomfort with frequent switching of regulation levels of such stabilizers.

Ferroresonant voltage stabilizers function by using a saturation choke. At a certain section of the volt-ampere characteristic of a saturation choke at insignificant voltage increments there are rather large current increments (saturation section). If such a choke is connected in series with a choke having a linear volt-ampere characteristic (linear choke)to ensure the saturation mode, then the voltage at the ends of the saturation choke will change insignificantly, when the voltage at the input of the resulting circuit changes. Due to high material intensity, the price of ferroresonant voltage stabilizers is high compared to the cost of electronic and electromechanical stabilizers. In addition, this type of voltage stabilizers introduces nonlinear distortions in the shape of the output voltage curve, does not allow operation in the mode with a disconnected load, and has a very narrow range of allowable deviations of the input line voltage.

Stabilizers with sub-magnetizable transformers perform voltage regulation for voltage correction by redistributing alternating magnetic fluxes in the structural elements of the transformer's specially arranged magnetic core.

A large material intensity is the main disadvantage; however, it is inferior to the material intensity of ferroresonant voltage stabilizers, especially at powers over 1 kW and about 1.5 times (by weight of materials) exceeds this indicator of a power autotransformer or transformer of the same power. In addition, there are distortions of the sinusoidal shape of the output voltage curve due to the use of nonlinear elements–magnetic shunts–in the magnetic conductors of transformers of these stabilizers.

Inverter three-phase voltage stabilizers use the technology of double conversion of electrical energy. They first rectify the input three-phase AC mains voltage and filter it to eliminate ripple and high-frequency noise. Then, using pulse width modulation technology (PWM technology), the DC voltage is converted (inverted) into a system of three alternating voltages with a sinusoidal curve and with initial phases shifted relative to each other by 120 degrees, according to Sadeque, Benzaquen and others [10]. These stabilizers have relatively low material consumption, high speed and accuracy of maintaining the specified voltage level, and smooth voltage regulation.

However, inverter voltage stabilizers are characterized by a number of disadvantages. The use of semiconductor electronic components such as diodes, insulated-gate bipolar transistors, and metal-oxide-semiconductor field-effect transistors in the design of inverter voltage regulators makes such stabilizers particularly sensitive in terms of reliability and performance to significant overvoltage in the grid and overloads, according to Singh, Tamrakar, Mewada, and Gupta [11].

Of all the types of voltage stabilizers discussed, electronic, electromechanical, and inverter voltage stabilizers are currently the most common. To select the best criterion from different criteria for a wide range of applications, knowledge of critical parameters for most electrical equipment is necessary.

First, an electric drive is used in industry and households, which consumes up to 50% of the country's electricity. A common feature of electric motors is the consumption of inrush current seven times higher than the rated current. Therefore, a stabilizer designed for a wide range of loads must be able to withstand overloads. Second, the smooth operation and service life of electrical equipment depends on the quality of functioning and efficiency of the stabilizer. The efficiency of many stabilizers is at a low level. In accordance with the above findings, we can conclude that the most optimal for many types of electrical loads are inverter voltage stabilizers, which have high efficiency, accurate voltage symmetry, and with properly selected power keys, even high reliability.

16.5 CONCLUSION

Electricity quality in 0.4 kV power grids is one of the indicators directly affecting the energy efficiency of power grids and enterprises. Non-compliance with power quality standards can lead to economic damage, increased power losses, reduced productivity, and other negative consequences. The use of modern electrical equipment and automation systems in industrial areas has increased the need for standardized power quality. These systems are sensitive to deviations in power quality, which makes it necessary to pay special attention to ensuring reliable and quality power supply. Current and voltage asymmetry in 0.4 kV networks can lead to various negative consequences, including increased power losses, reduced reliability of power supply, phase overloading, malfunctioning of electrical machines, voltage drop, and equipment failure. Technical means to reduce current and voltage asymmetry in 0.4 kV power grids are considered. Implementation of solutions to eliminate asymmetry is crucial to optimize network operation, minimize power losses, and ensure efficient operation of the industrial complex.

REFERENCES

1. M. V. Borodin and A. I. Psarev, "Estimation of the actual quality of electric power and analysis of the number of appeals regarding the non-compliance of electric power quality with the normative values", *Agrotechnics and Energy Supply*, vol. 17, pp. 54–62, 2017.
2. V. E. Bolshev, B. V. Papkov, A. V. Bukreev, and D. V. Konkin, "Algorithm of application of two transformers instead of one at transformer substations 10/0.4 kV", *Electrotechnologies and Electrical Equipment in Agroindustrial Complex*, vol. 71, issue 1, pp. 24–30, 2024.
3. I. V. Naumov, M. V. Shevchenko, and S. N. Voyakin, "Influence of voltage asymmetry in the 0.38 kV electrical grid on the characteristics of three-phase induction motor", In Proceedings: Actual Issues of Energy in the Agricultural Sector. Materials of the All-Russian Scientific-practical Conference with International Participation. Blagoveshchensk, pp. 121–138, 2020.
4. A. A. Abdullaev, H. S. Nematzhonov, and I. Z. Ibrohimov, "Determination of the asymmetry factor in electric grids", *Universum: Technical Sciences*, issue 12-6(105), pp. 42–45, 2022. https://doi.org/10.32743/UniTech.2022.105.12.14820.
5. R. Yu. Zakurdaev and I. E. Chernetskaya, "Features of current and voltage sensors application in the calculation of electric grid asymmetry", *Izvestia South-West State University*, vol. 23, pp. 160–169, 2019. https://doi.org/10.21869/2223-1560-2019-23-3-160-169.
6. V. E. Bolshev and A. V. Vinogradov, "Review of the methods to reduce power losses in electrical grids containing renewable energy sources", *Electrotechnologies and Electrical Equipment in Agroindustrial Complex*, vol. 70, issue 2(51), pp. 57–67, 2023. https://doi.org/10.22314/2658-4859-2023-70-2-57-67.
7. A. V. Churin and S. V. Sukyasov, "Methods of measurement of voltage asymmetry coefficients in a 0.38 kV grid", *Actual Issues of Agrarian Science*, issue 50, pp. 36–44, 2024. https://doi.org/10.51215/2411-6483-2024-50-36-44.
8. I. V. Naumov, E. S. Fedorinova, and M. A. Yakupova, "Influence of asymmetry of phase currents on the increase of fire hazard", *Actual Issues of*

Agrarian Science, issue 47, pp. 27–35, 2023. https://doi.org/10.51215/2411-6483-2023-(2)47-27-35.

9. W. Wang, X. Gao, B. Fan, X. Zeng, and G. Yao, "Faulty phase detection method under single-line-to-ground fault considering distributed parameters asymmetry and line impedance in distribution networks", *IEEE Transactions on Power Delivery*, issue 37.3, pp. 1513–1522, 2021.
10. F. Sadeque, J. Benzaquen, A. Adib, and B. Mirafzal, "Direct phase-angle detection for three-phase inverters in asymmetrical power grids", *IEEE Journal of Emerging and Selected Topics in Power Electronics*, issue 9(1), pp. 520–528, 2020.
11. N. Singh, S. Tamrakar, A. Mewada, and S. K. Gupta, *Artificial Intelligence Techniques in Power Systems Operations and Analysis* (1st ed.). Auerbach Publications, pp. 1–72, 2023. https://doi.org/10.1201/9781003301820

Chapter 17

Energy management systems using artificial intelligence

Future trends, challenges and solutions

Harsh Pratap Singh, Kedar Nath Singh, Lokendra Singh Songare, Nagendra Singh, and Neeraj Arya

17.1 INTRODUCTION

Beyond a reactive paradigm, this transforming power generates a proactive as well as adaptive energy management system (EMS) able to foresee, cope with, and learn from the complexities of energy dynamics. Solving the challenges with intermittent renewable energy sources depends totally on artificial intelligence (AI) being able to analyze massive amounts of data in real time [1–4]. AI-powered systems have great potential to drastically reduce greenhouse gas emissions by means of improved energy use and support of conservation habits. This conforms with world environmental targets and emphasizes the need to include AI in energy management strategies. The chapter addresses how AI alters the environment to support future energy management strategies. The major goals are to find such patterns and examine the prospective effects of fresh trends in AI applications for energy management. The study intends to enlighten decision-makers, legislators, and stakeholders on the shifting opportunities and approaching challenges. This extensive research opens the way for a more powerful and flexible energy landscape and helps the continuous debate on ecologically sustainable energy sources.

Starting in the late 20th century, the chapter highlights the development of energy management AI applications. Using machine learning methods, load forecasting, predictive maintenance, and grid optimization; significant events show the significant transformation in the power sector arising from distributed sources, renewable energy, and digital solutions. There are gaps and chances for future AI integration into energy management even with considerable progress. Among the challenges are the interpretability and explainability of AI models, which are critically required for stakeholder confidence and suitable application in crucial energy infrastructure. Large-scale power systems and smart grids still suffer from scalability issues. Still, future development chances abound—especially in decentralization, the Internet of Things (IoT), and advancements in reinforcement learning and

DOI: 10.1201/9781003516156-17

deep learning, which provide hopeful answers to challenging optimization problems in energy distribution and storage.

17.1.1 Background

Accurate and comprehensive energy usage information offered by EMS lets users reach long-term goals. They therefore greatly lower energy waste, so benefiting the user, the environment, and society at large. Published by the International Organization for Standardization (ISO), ISO 50001 [5] offers assistance to businesses using an EMS for the most effective energy consumption [6]. Claims that this criterion has evolved into a clear benchmark for the application of those systems as it helps companies show that they apply an efficient EMS considering the technological, socio-cultural, and financial particularities of every nation. Based on a system of inspections and indicator monitoring, ISO 50001 supports ongoing control of energy efficiency as part of the organizational culture shown in Figure 17.1. Feedback thus turns into a tool for learning that "improves the effectiveness of other information and advice in achieving better understanding and control of energy use" [7].

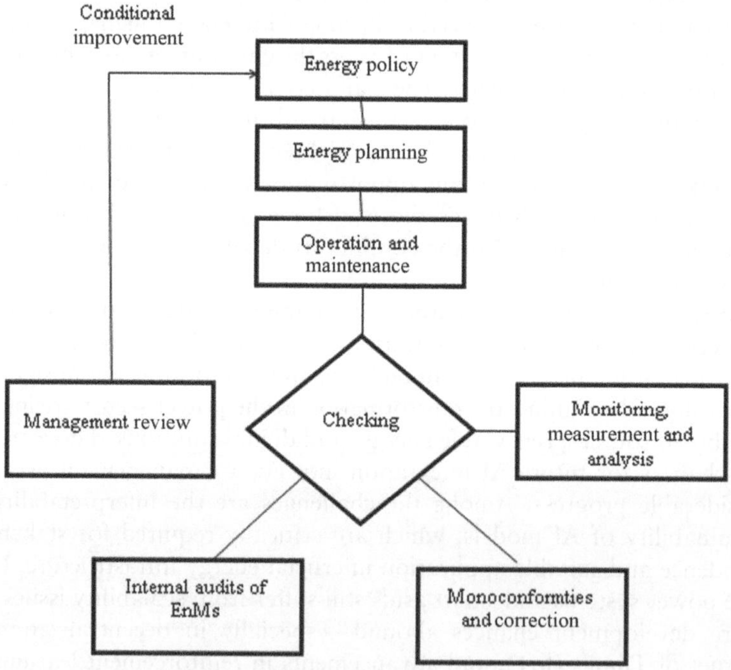

Figure 17.1 Continuous improvement on ISO 50001.

17.2 FUTURE TRENDS IN AI-BASED EMS

Several main trends are determining the direction of AI in energy management:

- *Machine learning for demand forecasting:* Effective resource allocation and grid stability depend on high-accuracy energy demand prediction. By means of past data on weather trends, consumer behavior, and industrial activity, AI algorithms can very precisely forecast demand patterns. This enables utilities to proactively change their distribution and generation plans, therefore reducing dependency on costly peak power plants and guaranteeing a seamless energy flow [8].
- AI-driven integration of renewable energy sources optimizes them because solar and wind energy sources are variable and intermittent and including them in the system offers special difficulties. AI may greatly enhance this integration by predicting renewable energy generation depending on weather forecasts and prior data. Figure 17.2 shows the AI trends in power systems. This makes it possible to apply dynamic grid management strategies and energy storage devices, therefore ensuring a more consistent and effective energy mix.
- Rising distributed energy resources (DERs) including electric cars and rooftop solar panels demand a decentralized approach to energy management employing AI agents. Agents driven by AI contained inside DERs can maximize personal energy consumption. Imagine a smart house with an AI agent managing energy use dependent on real-time power pricing, temperature, and appliance use patterns. This collective intelligence at the device level can considerably improve general grid efficiency [8].

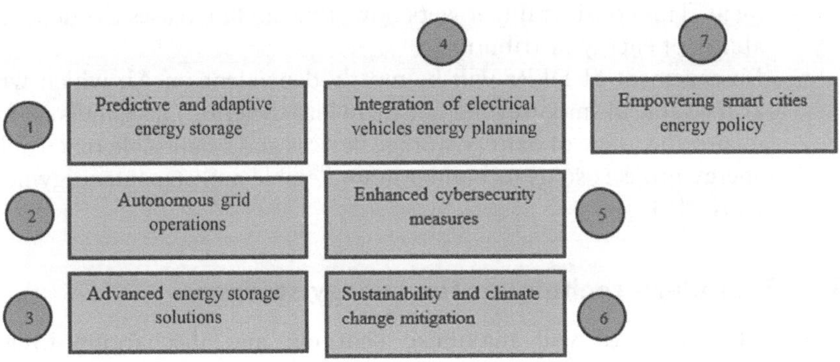

Figure 17.2 AI trends in energy sector.

- AI predictive maintenance for increased reliability major expenses and disturbance of the energy source might result from unplanned equipment breakdowns AI-driven predictive maintenance can search sensor data from power plants, transmission lines, and distribution networks to identify potential equipment faults before they begin. Early identification helps to enable early repairs and provides a more reliable electricity infrastructure, therefore lowering downtime.
- *Automated intelligence-driven cybersecurity concerning energy systems:* The energy sector is vulnerable to cyberattacks as reliance on linked devices and smart grids rises. Real-time identification and response to cyber threats are made possible by AI by means of network traffic pattern analysis, machine learning algorithms can identify anomalies suggestive of cyberattacks, so enabling quick mitigating techniques [9].

17.2.1 Predictive and adaptive energy systems

Predictive analytics driven by AI will become increasingly important in the future of AI in energy as they will help energy systems to more precisely forecast demand and supply. This entails projecting energy demand trends and, in light of environmental conditions, output from renewable sources such as solar and wind [9].

Real-time data will enable AI to help create adaptable energy systems that automatically change to meet evolving supply constraints and demand. This improves grid stability and efficiency, hence minimizing waste and besting resource utilization.

17.2.2 Autonomous grid operations

- AI developments will produce autonomous, self-healing energy grids able to identify, diagnose, and fix problems without human involvement. This considerably lowers downtime and increases the dependability of energy distribution.
- Management of DERs will be mostly dependent on AI, which will also be able to smoothly include them into the grid. This entails maximizing the usage of battery storage devices and small-scale renewable energy projects so as to build a more flexible and strong energy network [10].

17.2.3 Modern techniques for energy storage

- AI techniques will maximize charging and discharging cycles for battery storage systems, hence prolonging their lifetime and

efficiency. Powered by AI, predictive maintenance will aggressively find possible problems, therefore lowering the chance of failure and downtime.
- AI helps with more complex management of grid-scale energy storage, balancing supply and demand in real time and so smoothing out the fluctuation of renewable energy sources.

17.2.4 Incorporation of electric cars (EVs)

- As the acceptance of electric vehicles (EVs) rises, AI will help manage EV charging demand, minimize grid overload, and maximize charging schedules depending on grid capacity and renewable energy supply.
- AI will help to create vehicle-to-grid (V2G) technologies, in which electric cars can function as mobile energy storage units providing back-to-the-grid energy as needed. This two-directional energy flow improves grid stability and flexibility.

17.2.5 Improved security practices

AI-powered security systems: AI will enhance cybersecurity measures using powerful algorithms to detect and neutralize threats in real time, therefore securing important infrastructure from assaults as the energy sector gets ever more digital and linked.

17.2.6 Climate change prevention and sustainability

- AI will maximize energy generation, distribution, and consumption habits to lower carbon emissions, therefore assisting world efforts to slow down climate change.
- AI's capacity to control and project the output of renewable energy sources will help them to be included in the energy mix, therefore hastening the change to a sustainable energy future.

17.2.7 Encouraging smart cities

- AI will be essential in creating smart cities, optimizing energy use in buildings and public infrastructure, and so supporting effective urban planning to lower energy use and environmental impact.
- Innovation and promise abound in the direction of AI in the energy industry. These developments underline not only AI's power to transform energy management but also its part in promoting resilience, efficiency, and sustainability.

The integration of AI technologies into the energy sector will surely result in more intelligent, adaptable, and sustainable energy systems as they develop and mature, therefore preparing the ground for a day when energy is controlled in line with the demands of both people and the earth.

17.3 ISSUE AND CHALLENGES IN INTEGRATION OF AI WITH EMS

While integrating AI with EMS offers many possibilities, there are also certain difficulties. These are a few of the primary problems and obstacles [11].

17.3.1 Ethical concerns regarding AI

In AI, ethics describes arguments on numerous issues including social impact, perpetuation of bias, and privacy violations. These days, the challenge is responsibility, openness, and fairness AI systems apply to decisions taken. Algorithmic bias, which causes prejudice of particular groups, is one issue endangering the widening of inequality.

Moreover, AI should be used in sensitive areas like criminal justice and health that necessitate deeper consideration of ethical principles needed to obtain just results in a more concentrated way. Using AI for society's benefit hinges on balancing moral issues with technological advancement limiting risks and supporting ethical innovation.

17.3.2 Artificial intelligence

In AI, bias can be defined as the ability of machine learning algorithms to replicate and amplify already present preconceptions in the training set. Unethical and unfair outcomes resulting from this might significantly affect impoverished populations. AI bias mitigating calls for a methodical approach to data selection, preprocessing, and algorithm design to lower bias and enhance justice. Moreover, ongoing research and evaluation of AI systems help to detect and eliminate bias, thereby promoting fairness in the outcomes of every person by means of AI-based decision-making processes.

17.3.3 AI integration

Integration of AI is the process of adding AI systems into products and services to increase efficiency and automation. These points to identifying relevant application cases, customizing AI models to meet particular contexts, and ensuring system fit-ability. The integration procedure demands AI experts and domain experts to cooperate to refine their solutions to satisfy organizational needs. Among the challenges were related to change management, human training requirements, and data interoperability.

17.3.4 Digital control in computing

Developing and using AI models—especially those involving computations and big datasets—dependent on computational power mostly rely on its degree. Demand for high-performance computing devices such as graphics processing units (GPUs), tensor processing units (TPUs), and others results from the growing complexity of AI algorithms. Among the challenges lie scalability, energy consumption, and pricing. Early stages of research also offer viable solutions from hardware architectural developments including neuromorphic computing and quantum computing.

Furthermore, competent to overcome computational limits are distributed computation and cloud services. Maintaining limited resources, the development of AI possibilities depends on efficient and sustainable management of computing needs.

17.3.5 Personal privacy and security

Data security and privacy are the main issues around AI systems as their operation and training depend on enormous amounts of data. If one intends to stop leaks, breaches, and abuse, one must also assure data security, availability, and integrity. Following data protection laws such as the General Data Protection Regulation (GDPR) and California Consumer Privacy Act (CCPA) requires a company to have access limits, encryption, and auditing authority [11].

Moreover, using privacy-preserving methods such as differential privacy and federated learning helps to minimize privacy issues and keep data utility. Open data platforms and ethical data handling methods help users create confidence in AI systems and responsible data management.

17.3.6 Artificial intelligence legal issues

Legal questions surrounding AI generally cover liability to intellectual property rights and regulatory compliance. Particularly in a malfunctioning system or an accident caused by an autonomous AI system, an AI-based decision-maker engaged becomes clearly responsible. Legal questions about copyright arise from the ownership of the produced work by AI and its algorithms.

Moreover, the regulatory systems—usually the last to change with the times—have to be observed to prevent legal obligations and hazards. Dealing with these problems calls for a mix of legal experts, legislators, and technological professionals cooperating to produce unambiguous norms and policies balancing innovation with responsibility and safeguarding of rights for stakeholders.

17.3.7 AI visibility

Transparency in AI relates to decision-making techniques and algorithms. Developing credibility, trust, responsibility, and user confidence in AI systems depends on openness. Transparency is defined as an element of how AI models operate and what they do including inputs, outputs, and fundamental logic. Explainable artificial intelligence (XAI) is one of the methods meant to provide clearly intelligible insights into complicated AI systems [12].

Transparency would also be encouraged by well-defined performance measures, model training techniques, and data source documentation. Promoting transparency helps companies to display ethical AI methods, handle bias, and let consumers make the correct decisions depending on AI-derived results.

17.3.8 Restricted understanding of AI

One of the main problems influencing informed decision-making, acceptance, and control among the general people is limited knowledge. Misconceptions and misinterpretations of AI's capabilities and limitations proliferate, therefore hindering its responsible usage and advancement. To guarantee public understanding of AI principles, use cases, and possible effects, effective methods should be created and carried out for public awareness campaigns including education.

Moreover, allowing users to access easily available information and training chances will help them to apply AI technology more wisely. Bridging the knowledge gap through interdisciplinary collaboration, community involvement, and outreach is how society will get AI that can be productive without at the same time dealing with the ethical, societal, and economic issues of AI.

17.3.9 Insufficient AI explainability

The lack of AI explainability is the difficulty in knowing and figuring out how AI systems get to their conclusions or suggestions. This lack of openness makes individuals suspicious and responsible about important uses such as banking and healthcare. By offering insights into the reasoning of AI algorithms, the AI techniques are meant to solve this problem. Visualizing models and feature analysis help consumers understand AI outcomes. Although the explainability problem is still a difficult one, that fix does not influence the model performance [13].

17.4 APPLICATION AREAS OF AI IN EMS

There are many and always-changing possible uses of AI in energy control. These are many notable instances:

- AI systems can dynamically change voltage levels, reroute electricity depending on congestion, and manage distributed energy supplies thereby optimizing energy flow inside the grid.
- AI-powered systems can encourage users to change their energy usage patterns at periods of maximum demand, therefore lessening dependency on costly power plants.
- AI can examine building use data to find chances for energy efficiency improvements—that is, for optimizing heating, ventilation, and air conditioning (HVAC) systems.
- AI-powered systems can enable users to track their energy consumption, understand their environmental impact, and make educated decisions to lower their carbon footprint.
- AI systems can examine market trends and forecast energy prices, thereby helping utilities to make the best decisions about energy procurement and trading [14].

17.5 PROPOSED SOLUTIONS FOR THE INTEGRATION OF AI WITH EMS

17.5.1 Quality in data management

The sheer volume of data from many sources—including smart meters, sensors, and satellite images—can be daunting. Inconsistencies, errors, and lack of standardizing increase the problem, therefore reducing the dependability of data for important decisions [15]. Solution for AI: By automating data cleaning and standardizing processes, advanced AI algorithms and machine learning methods help to guarantee high-quality datasets. These techniques can also reveal patterns from unstructured data, therefore opening fresh chances for the energy economy.

17.5.2 Compatibility with existing systems

Challenge: The dependence of the energy sector on outdated, rigid infrastructure can greatly limit the integration and scalability of AI solutions, therefore impeding invention [15].

AI middleware solutions can help to link new technologies with out-of-date systems. These solutions let data formats and protocols translate, therefore enabling perfect communication and integration without requiring significant infrastructure changes.

AI-enhanced cybersecurity systems can provide real-time monitoring and threat detection by way of predictive analytics to identify possible flaws and assaults before they materialize. AI also helps to minimize the time and money required to manage invasions by automating responses to security events.

17.5.3 Privacy and controlling concerns

Compliance with evolving norms and standards for data protection and energy management can be frightening, especially with respect to emerging technologies like AI. Legal criteria can be included in AI systems' algorithms, hence guiding compliance. Automated compliance solutions driven by AI can help energy companies traverse difficult regulatory surroundings and ensure that operations remain within legal constraints.

The move toward AI-driven energy systems requires a workforce capable of both AI and energy management, a combination that is not now very common. Designed in tandem with academic institutions and technology companies, comprehensive education and training programs might equip current and future employees with the necessary skills utilizing AI. AI can also help this educational process by means of simulations and customized learning opportunities.

17.5.4 First costs are substantial

Smaller operators could be deterred from embracing the significant cost involved in implementing AI—including the technology itself as well as the infrastructure to support it.

As-a-service models and cloud-based AI technologies offer reasonably cost-effective scalable solutions that might lower the entrance barrier. These solutions allow companies to pay for AI capabilities as they use them, therefore reducing upfront costs and financial risk [16]. Including AI in the energy sector presents a clear path to not only fix present issues but also grab new opportunities for security, sustainability, and efficiency.

By addressing data quality, system integration, cybersecurity, regulatory compliance, workforce development, and cost, AI opens the road for a more resilient and flexible energy infrastructure. As the sector grows, AI will definitely serve several purposes, encouraging innovation and enabling global society to move toward a more sustainable energy future [17, 18].

17.6 CONCLUSION

A major first toward maximizing energy use, improving efficiency, and supporting sustainability is the integration of AI with energy management systems. This convergence of technology portends a time when energy management will be more predictive, flexible, and sensitive to both customer requirements and environmental concerns. Successful integration of AI with EMS will define energy management going forward. Although the path is full of difficulties, the possible advantages in terms of sustainability, economy, and cost savings are rather great. Through creative solutions addressing technical, financial, organizational, and legal obstacles, we can fully utilize AI to build a smarter, more sustainable energy future.

REFERENCES

1. Kubassova, O.; Shaikh, F.; Melus, C.; Mahler, M. "History, Current Status, and Future Directions of Artificial Intelligence," In *Precision Medicine and Artificial Intelligence*, Elsevier, Amsterdam, The Netherlands, 2021, pp. 1–38. ISBN 978-0-12-820239-5.
2. Turing, A.M. "Computing Machinery and Intelligence," In *Parsing the Turing Test: Philosophical and Methodological Issues in the Quest for the Thinking Computer*, Epstein, R., Roberts, G., Beber, G., Eds., Springer, Dordrecht, The Netherlands, 2009, pp. 23–65. ISBN 978-1-4020-6710-5.
3. Borkar, E.; Singh, N. "Power Quality Enhancement by PV-UPQC for Non-Linear Load," In *Artificial Intelligence Techniques in Power Systems Operations and Analysis*, Taylor& Francis, July 2023. ISBN. 978-1-0033-0182-0.
4. Odagiri, H.; Nakamura, Y.; Shibuya, M. "Research Consortia as a Vehicle for Basic Research: The Case of a Fifth Generation Computer Project in Japan," *Research Policy*, 26, 191–207, 2009. https://doi.org/10.1016/S0048-7333(97)00008-5.
5. Singh, N. "Application of Fiber Optics for the Protection and Control of Power Systems," In *Optical to Terahertz Engineering. Progress in Optical Science and Photonics*, Springer, Singapore, 2023, vol. 23, pp. 101–120. https://doi.org/10.1007/978-981-99-0228-6_7
6. Sonika, R.; Singh, N. "Comparative Analysis of 36, 48, 60 Pulse AC-DC Controlled Multipulse Converter for Harmonic Mitigation," *International Journal of Advanced Research in Computer Engineering & Technology (IJARCET)*, 3(4), 1159–1163, April 2014. ISSN: 2278-1323.
7. Veena, R.S.; Khambra, G.; Sayal, R.; Singh, N.; Bala Dhandayuthapani, V.; Sar, S.K. "An Energy Efficient IPv6 Packet Delivery Outline for Industrial IoT Over G.9959 Protocol Based Wireless Sensor Network (WSN)," *SN Computer Science*, 5, 813, 2024. https://doi.org/10.1007/s42979-024-03195-4
8. Singh, N.; Kumar, Y. "Multi Objective Economic Load Dispatch Problem Solved by New PSO," *Advances in Electrical Engineering*, 2015, 2356–6655, 2015. https://doi.org/10.1155/2015/536040
9. Nighoskar, A.; Chaurasia, P.; Singh, N. "Advancing Nuclear Energy Forecasting: Exploring Regression Modeling Techniques for Improved Accuracy," *Nuclear Engineering and Technology*, 57, 103144, 2025. https://doi.org/10.1016/j.net.2024.08.013
10. Krishna, S.R.; Kumar, R.; Gaurav, A.; Singh, N.; Sharma, M.; Almas, S.K. *Intelligent Adaptive Systems and Methods Thereof for Household Energy*

Control and Management, In 2024 IEEE 3rd International Conference on Electrical Power and Energy Systems (ICEPES), Bhopal, India, 2024, pp. 1–3. https://doi.org/10.1109/ICEPES60647.2024.10653541

11. Singh, N.; Chakrabarti, T.; Chakrabarti, P.; Margala, M.; Gupta, A.; Krishnan, S.B.; Unhelkar, B. "A New PSO Technique Used for the Optimization of Multiobjective Economic Emission Dispatch," *Electronics*, 12(13), 2960, 2023. https://doi.org/10.3390/electronics12132960
12. Singh, N.; Chakrabarti, T.; Chakrabarti, P.; Margala, M.; Gupta, A.; Praveen, S.P.; Krishnan, S.B.; Unhelkar, B. "Novel Heuristic Optimization Technique to Solve Economic Load Dispatch and Economic Emission Load Dispatch Problems," *Electronics*, 12(13), 2921, 2023. https://doi.org/10.3390/electronics12132921
13. Singh, N.; Nighoskar, A.; Ram Krishna, S. "Economic Load Dispatch with Valve Loading Effects Optimize by BBMO," *International Journal of Innovative Technology and Exploring Engineering (IJITEE)*, 8(7), 104–107, May 2019.
14. Al-Abassi, A.; Karimipour, H.; Dehghantanha, A.; Parizi, R.M. "An Ensemble Deep Learning-based Cyber-attack Detection in Industrial Control System," *IEEE Access*, 8, 83965–83973, 2020.
15. Singh, N.; Praveen Kumar, M.; Santosh Kumar, B. "Effect of Valve Loading on the Thermal Power Economic Load Dispatch Using New Elephant Herding Optimization," *International Journal of Recent Technology and Engineering (IJRTE)*, 7(6S4), 345–349, April 2019. ISSN: 2277-3878.
16. Jain, A.; Singh, N. "Shunt Active Power Filter Controlled for Mitigation of Harmonics," *International Journal of Advanced Technology & Engineering Research (IJATER)*, 5(6), 23–27, November 2015.
17. Shi, J.; Gao, Y.; Gu, D.; Li, Y.; Chen, K. "A Novel Approach to Detect Electricity Theft Based on Conv-attentional Transformer Neural Network," *International Journal of Electrical Power & Energy Systems*, 145, 108642, 2023.
18. Sun, Y.; Lee, J.; Kim, S.; Seon, J.; Lee, S.; Kyeong, C.; Kim, J. "Energy Theft Detection Model Based on VAE-GAN for Imbalanced Dataset," *Energies*, 16, 1109, 2023.

Chapter 18

Replacement of network charging stations for autonomic stations for the needs of agricultural machinery

Anton Sporov and Leonid Yuferev

18.1 INTRODUCTION

Many scientists and electric vehicle manufacturers are working on the development of charging station designs for charging batteries from renewable energy sources. For example, some scientists [1] propose a mobile solar power station for charging low-power batteries. This mobile solar power station includes four photovoltaic (PV) modules and a voltage converter with the capability to control the charge and protect the batteries [2]. The drawback of this station is the low charge current output, which limits its use for charging electric vehicle batteries. Meanwhile, the task of creating battery-charging stations for electric vehicles is relevant for agricultural enterprises and farms, where such transport can be used for monitoring product quality, controlling technological processes, and other purposes. The power of such electric vehicles can reach upto 10 kW or more, so the station must have the appropriate technical characteristics.

Another issue in charging stations is the placement of the charging station. It is necessary to consider the optimal location of the solar charging station so that electric vehicles using this setup have easy access to it. Solar batteries (SBs) should be placed in open spaces. The location of the additional battery set should be within a protected area. Therefore, it is recommended to install the solar power station in a garage. The power station should consist of SBs installed on the southern slope of the garage roof, a switch box, a charge controller, and connectors located inside the building. SBs should be selected based on the geographical latitude of the farm's location.

SBs must have the necessary power to charge the batteries. The power station should consist of SBs installed on the southern slope of the garage roof, a switch box, a charge controller, and connectors located inside the building.

DOI: 10.1201/9781003516156-18

18.2 ORGANIZATION OF ELECTRICALLY DRIVEN AGRICULTURAL TRANSPORT AND SOLAR BATTERIES

An agricultural vehicle (AV) with an electric drive contains several batteries that need to be periodically charged. The charging period depends on the operating mode of the electric vehicle. Due to the need to monitor technological production, specific operating modes for the enterprise must be developed. Generally, most work is done during the day, which is also when the solar charging station operates. Therefore, to ensure the AV is operational during work hours, at least two sets of batteries are needed. One set is installed in the electric vehicle, and the other set is being charged. The sets are swapped the next day [3].

The SBs should be installed in open areas without any shading from trees or buildings. It is important to carefully plan their positioning. The tilt angle should be adjusted according to the farm's latitude, and the azimuth should be set to 180°. For example, at the equator, the energy of solar radiation is much higher than at the North Pole, due to the solar rays passing through less atmospheric mass. Therefore, at the equator, the SB tilt angle will be almost horizontal [4].

Installing the SBs on the garage roof has its advantages, as the roof slope angle in central Russia almost matches the required SB tilt angle. More precise installation is achieved using brackets that provide the pre-calculated optimal angle and allow manual adjustment of the SB tilt angle twice a year – in winter and summer, thereby increasing the conversion efficiency [4]. The tilt angles of the solar panels at different times of the year are shown in Figure 18.1.

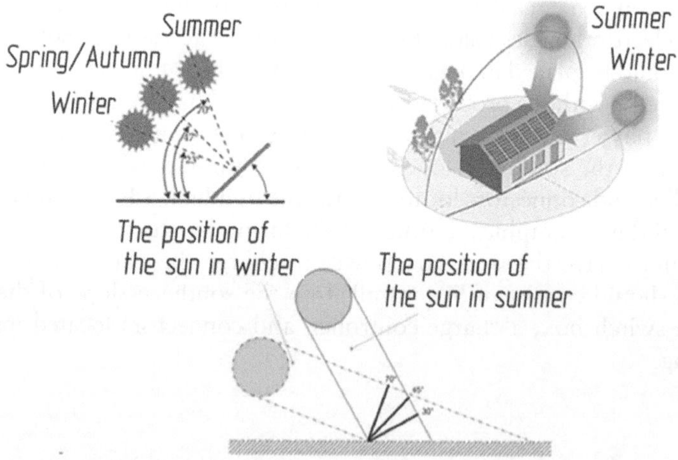

Figure 18.1 Angle of declination of solar panels at different times of the year season.

18.3 CALCULATING THE CAPACITY OF A SOLAR CHARGING STATION AND ITS LOCATION

A farmer's solar power station has been proposed for setting up at an agricultural enterprise (farm). The batteries of the agricultural electric vehicle are charged using solar panels installed on the roof of this power station. The placement of the elements of the farmer's solar power station with the battery charging system is shown in Figure 18.2. The arrangement of the elements of the farmer's solar power station is illustrated in Figure 18.2.

Figure 18.2 shows the following components: 1, SBs; 2, switch box; 3, solar charge controller; and 4, battery. The basic electrical schematic for charging batteries from the SBs is presented in Figure 18.3.

Figure 18.3 shows FC1 – SBs; Rsh – shunt resistor; R1 – resistor; VD1 – diode; C1 capacitor (10,000 μF); G1 – battery.

The solar charging station must provide the necessary charge level even under diffuse sunlight conditions [5]. The solar panel is used only to charge the batteries, which should be done as needed based on the battery discharge. Under unfavorable conditions, charging may continue for the entire daylight period. At night, there is no need to disconnect the solar panels from the batteries as they will be automatically disconnected by diode VD1 [6].

The standard charging time for lead-acid and alkaline batteries is 10% of their capacity, or 10 hours. The fastest charging batteries are lithium-ion,

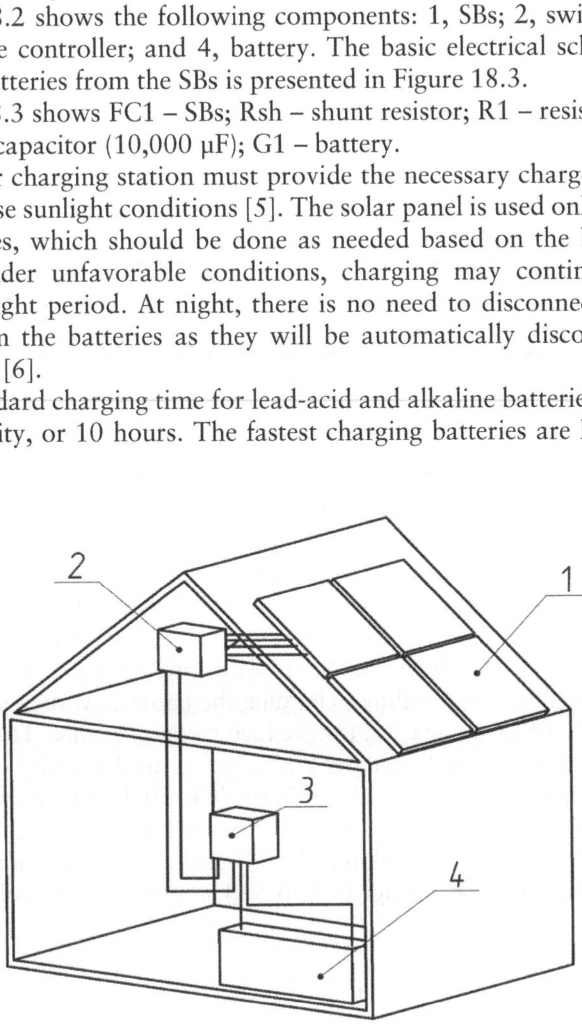

Figure 18.2 Placement of solar power plant elements with battery charging system where: 1, solar panels; 2, junction box; 3, solar charge controller; and 4, accumulator.

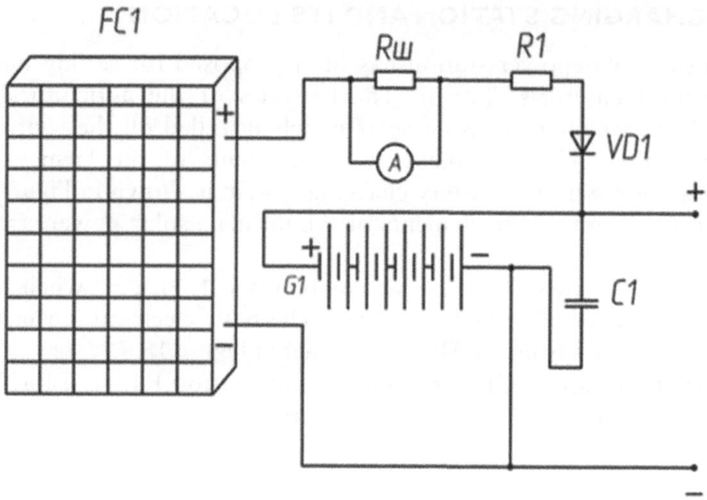

Figure 18.3 Electrical diagram of connection of solar panels to batteries.

which take about 4 hours to fully charge. Under unfavorable conditions, charging may continue throughout the day. At night, solar panels do not need to be disconnected from the batteries as they will be automatically disconnected by diode VD1 [7].

For the most efficient battery charging, the SBs should be configured to generate the maximum current, achievable when the SB voltage is higher than the battery voltage. The SBs should be connected to the batteries through diode VD1, as shown in Figure 18.1. This prevents the batteries from discharging through the SB internal resistance when solar radiation is insufficient, and the SB voltage drops below the battery voltage. Capacitor C1 is necessary if the batteries are charged in buffer mode [8]. Resistor R1 is used to limit the constant current. Buffer mode allows charging the batteries with a current equal to 20%–25% of their capacity, thus reducing charging time. However, such modes should not be used regularly but only in critical situations [8].

Experimental studies of the electric vehicle's capabilities showed that the installed battery provided a range of 60 km on a single charge cycle, which is sufficient for a farmer's working day. The capacity of the installed battery is 9 Ah with a nominal voltage of 540 V. The power of the solar charging station is:

$$P_{se} = \left(U_n \times I_z \times k\right) \div 1 = 730\ W \tag{18.1}$$

where P_{se} is the calculated power of the solar charging station, W;
U_n is nominal battery voltage,V;
I_z is the charge current, taken as 1 A;
k is the optical loss coefficient [7];
t is time, 1 hour.

Based on the power of the solar charging station, SBs are selected according to the following algorithm: The number of SBs is determined based on the voltage of the battery units:

$$n = U_{AB} \div U_A = 540 \div 13.5 = 40 \ pcs. \tag{18.2}$$

Where n quantity of batteries in the battery pack;
U_{AB} is the voltage of the battery pack is 540 V.
U_A is the voltage of a fully charged battery is 13.5 V.

Select solar panels with a power rating of 250 W. Choose SBs with a voltage of 24 V, and the required number of solar panels needs to be calculated and matched with their power rating:

$$N_p = U_{AB} \div U_{SB} = 540 \div 24 = 22.5 \approx 23 \ pcs., \tag{18.3}$$

Where U_{SB} – is the nominal voltage of one solar panel.
Calculation of the capacity of one solar charging power station is made, pcs.:

$$P_{pan} = P_{se} \div n_{pan} = 730 \div 23 = 30 \ W. \tag{18.4}$$

where n_{pan} is number of SB, pcs.;
P_{se} is capacity of solar charging power station, W;
P_p is power of one SB, W.

We choose the solar panels nearest by power:

Delta SM 30-24 solar panel is selected, 30 W and 24 V nominal voltage.
Schneider Electric Conext MPPT 80 A 600 V solar charge controller is selected.
The selected SBs are connected in series on the garage roof as shown in Figure 18.2.

18.4 PARAMETER ESTIMATION OF SB SYSTEM COMPONENTS INTEGRATED IN A RURAL INFRASTRUCTURE BUILDING

The technical specifications of SB systems integrated into rural infrastructure buildings (IFES) and their components should include specific parameters that differ from traditional non-integrated systems. It is important to consider the parameters and characteristics of the SB array of the IFES, including modules integrated with building structures (IFM), mounting structures for the installation of the IFM, and fastening components and parts. For the remaining IFES components, the requirements remain the same as for traditional systems and are evaluated in a similar manner [9, 10]. Figure 18.4 shows the energy flow distribution of the AV.

18.4.1 Mandatory parameters and characteristics of the PV battery

The differences in the performance of IFES arise due to the need to take into account the parameters of the building structure, which requires additional testing. As the IFM and the IFB (integrated solar battery) fulfill the functions of both SB equipment and building elements, traditional tests are modified. The IFES specification should include information on the type of building into which the system is integrated, a description of the building structures into which the IFB and IFM are integrated, IFM classification data, main

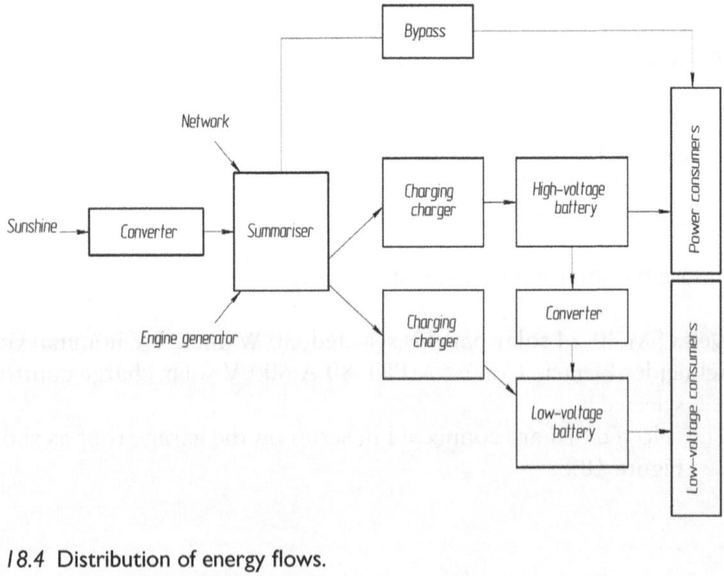

Figure 18.4 Distribution of energy flows.

characteristics and parameters, operating conditions, and storage and transport requirements [11].

Mandatory IFES parameters include rated inverter power, rated IFB power under standard conditions, performance under design solar radiation, design solar radiation arrival, rated direct current (DC) voltage, maximum possible IFB voltage, inverter output voltage and current, number of alternating current (AC) phases, DC voltage at load input, IFB volt–ampere characteristics under different light and temperature conditions, system and component efficiency, and IFM power degradation. These parameters and characteristics should consider the installation characteristics of the FMI and their impact on performance and property retention [12].

The set of construction characteristics of IFBs and IFMs and the methods for determining them depend on the structure with which they are integrated. The mandatory construction characteristics of IFBs and IFMs depend on the type of construction and IFM group are listed below.

Mandatory construction characteristics of IFBs and IFMs include the following:

1. Materials and structural strength.
2. Thermal insulation properties.
3. Resistance to weather conditions.
4. Fire resistance.
5. Sound insulation level.
6. Overall dimensions and weight.
7. Methods of attachment and integration into the building.
8. Resistance to wind and mechanical loads.
9. Durability and service life.
10. Compliance with building codes and standards.

18.4.2 Classification of integrated PV modules

The mandatory parameters and their evaluation depend on the type of construction and features of the FMI. The classification of an FMI takes into account the influence of external structures and the location of the FMI relative to the building and radiation. For the estimation of the parameters of the IFES components, it is important to consider two ways of dividing IFMs by design and two ways of classifying the production of IFMs [13].

IFMs are divided into two types:

1. Translucent, with a single glass-coated photovoltaic (PV)part, metal, or polymer-waterproofing sheet on the backside.
2. IFM with the PV part on the front side (group I), inside the structure (group II), or on the back side (group III). IFMs of groups I and II can be of any of the first two variants, and group III can be of the

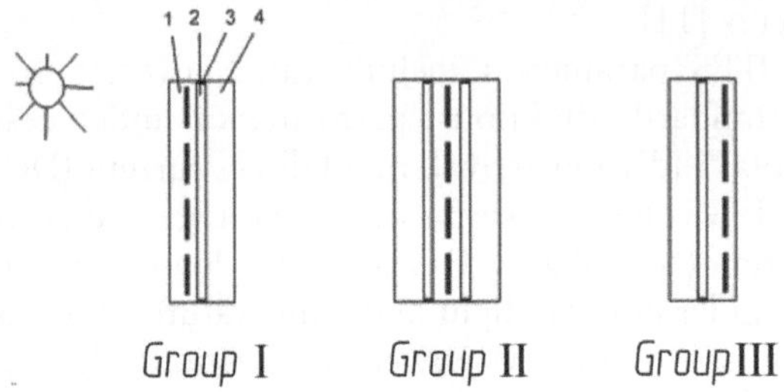

Figure 18.5 Schemes of constructions of groups I–III on the example of insulating glass units: 1, PV; 2, airgap; 3, sealant; and 4, glass.

third variant only. Schemes of constructions of group's I–III on the example of double-glazed windows are shown in Figure 18.5.

According to the organization of production, FMIs are divided into the following:

1. Mass-produced IFMs of one variant of final design: These are mass-produced IFMs with basic design, where dimensions and density of filling with PV elements are changed according to design requirements; IFMs for individual projects.
2. IFMs where different manufacturers produce the PV part and the final structure: These include IFMs where the same manufacturer in the same technological process produces the PV part and the final structure.

18.4.3 Evaluation of the performance of a PV array integrated into the roof of a combine harvester

Initial data:

1. PV modules should be placed on the whole surface of the harvester roof, consisting of two parts of the area A1 and A2 (see Figure 18.5).
2. The territories of possible utilization of the harvester are highlighted in Figure 18.6. The regions of the Volga region and the south of the European part of the Russian Federation are of the greatest interest.
3. The position of the modules is horizontal.
4. Working period July–September, working time 9–16 hours.
5. Operating conditions: under cloudless sky (maximum estimated arrival) and under standard cloudiness (minimum estimated arrival).

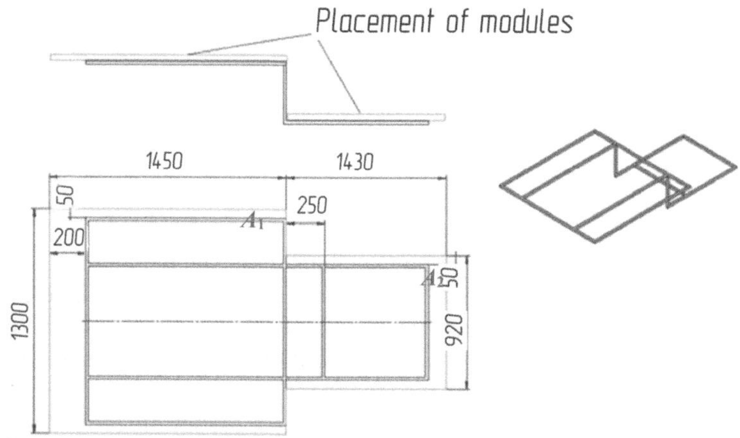

Figure 18.6 Sketch of the combine harvester roof.

Table 18.1 Maximum daily output of PV integrated on the roof of the combine harvester in July, kWh

Indicative point of the area	*July*	*August*	*September*
Samara	3,486	2,790	1,641
Voronezh	3,232	2,513	1,602
Kazan	3,237	2,541	1,619
Rostov-on-Don	3,317	2,926	2,179
Kostroma	3,203	2,507	1,551
Volgograd	3,764	3,005	1,602
Sochi	3,877	3,430	2,077
Astrakhan	3,560	3,237	2,043

The maximum yield is provided when the surface area on the combined roof is equal to the area of the working surfaces of the modules (losses of energy illumination only on the frame of modules along the perimeter of each part when one module is placed on each part). If several modules are placed on each part closely together without gaps between them, the yield will be slightly lower (losses per module frame). The area over which radiation converted into electricity is received is defined as

$$A = \left(A_1 - A_{frames1}\right) + \left(A_2 - A_{frames2}\right). \tag{18.5}$$

Before selecting the modules, the calculation is carried out for the maximum area of the working surfaces of the modules. The efficiency of the modules is assumed to be 20%. Standard frame width of crystalline silicon-based modules is 4 mm and A = 2.8298m^2.

Estimated values of maximum possible monthly and daily output in July–September for Russian cities were obtained in the form of graphs and tables (61 indicative points), and values of maximum daily output for the most significant indicative points of the area are given in Table 18.1.

18.5 CONCLUSION

A fragment of a farmer's solar power plant with SBs for charging electric vehicle batteries was developed. The question of SB tilt angle was investigated. The electrical scheme was built, the power of SB was calculated, the power of SBs was selected, and the required amount of SBs for charging the battery was calculated.

The parameters of the components of the PV system integrated into the rural infrastructure building were evaluated. The technical characteristics of the PV system integrated into the building of rural infrastructure and sets of mandatory parameters of its components were determined. The classification of the IFM was completed. An evaluation of the yield of a PV battery installed on the roof of a combine harvester was performed.

Power plants for charging mobile vehicles using solar panels can significantly reduce the load on the standard electricity grid, as they allow vehicles to be charged autonomously using renewable energy. This reduces the consumption of electricity from the grid, especially during peak hours, which helps to stabilize the grid and reduce the risk of overloading. However, if solar power plants are integrated into the overall grid, they can also contribute excess electricity to the grid, which requires effective control and balancing to maintain grid stability, and reliability. In the long term, the widespread use of such systems may stimulate the development and modernization of electrical infrastructure to better integrate renewable energy sources.

REFERENCES

1. M. Amir, Zaheeruddin, A. Haque, Intelligent energy management scheme-based coordinated control for reducing peak load in grid-connected photovoltaic-powered electric vehicle charging stations, *IET Generation, Transmission & Distribution*, 2023. doi:10.1049/gtd2.12772
2. I. V. Belokonov, A. V. Kramlikh, I. A. Lomaka, P. N. Nikolaev, Reconstruction of a spacecraft's attitude motion using the data on the current collected from solar panels, *Journal of Computer and Systems Sciences International*, Vol. 58, no. 2, pp. 286–296, 2019. doi:10.1134/S1064230719020047
3. A. P. Sporov, D. Y. Pisarev, A. S. Parakhnich, Electric drive of an agricultural vehicle, *Agricultural Machines and Technologies*, Vol. 15, no. 3, pp. 48–54, 2021. doi:10.22314/2073-7599-2021-15-3-48-54

4. C. V. Mitrofanov, Selection of the optimal angle of inclination of solar panels for their placement in an arbitrary region, *Bulletin of South Ural State University. Series: Energy*, Vol. 23, no 1, pp. 5–11, 2023. doi:10.14529/power230101
5. Yu. Zhu, J. Jin, X. Wang, An FBG sensor placement method to achieve long fatigue life and high measurement accuracy in deformation sensing of satellite solar panels, *IEEE Sensors Journal*, Vol. 24, no. 1, pp. 318–328, 2024. doi:10.1109/jsen.2023.3335256
6. A. P. Sporov, V. A. Gusarov, D. Y. Y. Pisarev, Trends in the development of agricultural electrified mobile equipment, *Electrotechnologies and Electrical Equipment in Agroindustrial Complex*, Vol. 69, no. 1(46), pp. 3–8, 2022. doi:10.22314/2658-4859-2022-69-1-3-8.
7. Yu. Yuferev, A. P. Sporov, Methodology of calculation of battery selection for mobile vehicle with electric drive, *Electrotechnologies and Electrical Equipment in Agroindustrial Complex*, Vol. 70, no. 3(52), pp. 109–114, 2023. doi:10.22314/2658-4859-2023-70-3-109-114
8. V. A. Gusarov, A. P. Sporov, Farmer solar charging power station for electric transport, *Agrotechnics and Energy Supply*, Vol. 2, no. 27, pp. 54–61, 2020.
9. R. S. Veena, G. Khambra, R. Sayal, N. Singh, B. Dhandayuthapani, S. Kumar Sar. An energy efficient IPv6 packet delivery outline for industrial IoT over G.9959 protocol based wireless sensor network (WSN). *SN Computer Science*, Vol. 5, p. 813, 2024. https://doi.org/10.1007/s42979-024-03195-4
10. E. V. R. Gómez, C. E. B. Vargas, K. L. Arias, D. C. H. Gutierrez, Implementation of solar panels and photovoltaic systems as an alternative for efficient energy saving at Universidad Nacional Abierta y a Distancia-UNAD, *Journal of Sustainability Perspectives*, Vol. 2, 2022. doi:10.14710/jsp.2022.15535
11. N. Singh, S. Tamrakar, A. Mewada, S. K. Gupta, *Artificial Intelligence Techniques in Power Systems Operations and Analysis* (1st ed., pp. 1–72), 2023, Auerbach Publications. https://doi.org/10.1201/9781003301820
12. N. Singh, A. K. Sharma, I. Sarkar, S. Prabhu, K. Chadaga, IoT-based greenhouse technologies for enhanced crop production: A comprehensive study of monitoring, control, and communication techniques, *Systems Science & Control Engineering*, Vol. 12, no. 1, 2024. https://doi.org/10.1080/21642583.2024.2306825
13. O. V. Shepovalova, Mandatory characteristics and parameters of photoelectric systems, arrays and modules and methods of their determining, *Energy Procedia*, Vol. 157, pp. 1434–1444, 2019. doi:10.1016/j.egypro.2018.11.308

Index

Index